Fariba Fahroo, Le Yi Wang, and George Yin (Eds.)

Recent Advances in Research on Unmanned Aerial Vehicles

Springer

Editors
Fariba Fahroo
AFOSR/RSL
Computational Mathematics
Arlington
Virginia
USA

George Yin
Department of Mathematics
Wayne State University
Detroit
Michigan
USA

Le Yi Wang
ECE Dept.
Wayne State University
Detroit
Michigan
USA

ISSN 0170-8643
ISSN 1610-7411 (electronic)
ISBN 978-3-642-37693-1
ISBN 978-3-642-37694-8 (eBook)
DOI 10.1007/978-3-642-37694-8
Springer Heidelberg New York Dordrecht London

Library of Congress Control Number: 2013935113

Printed on acid-free paper

Springer is part of Springer Science+Business Media (www.springer.com)

Lecture Notes in Control and Information Sciences 444

For further volumes:
http://www.springer.com/series/642

Preface

Unmanned aerial vehicles (UAVs), also known as unmanned aircraft systems (UAS) or drones, have seen an unprecedented growth recently both in military missions and in new developments toward civil and commercial applications. Although remotely operated small UAVs are commercially available and their successful operations in reconnaissance, combat operation, logistic support have been well recognized, autonomous control of UAVs as a team for a common mission remains a very challenging task.

A team of launched and coordinated UAVs requires advanced technologies in sensing, communication, computing, and control to improve their intelligence and robustness towards autonomous operations. At present, capabilities for command, control, communications,computing, intelligence, surveillance, and reconnaissance of networked UAVs are inadequate which restrict potential applications of UAVs. To enhance reliability, robustness, and mission capability of a team of UAVs, a system-oriented and holistic approach is desirable in which all components and subsystems are considered in terms of their roles in and impact on the entire system.

This volume aims to summarize the recent progress, identify challenges and opportunities, and develop new methodologies and systems on coordinated UAV control. A group of experts working in this area have contributed to this volume in several diversified but related aspects of autonomous control of networked UAVs. Their papers introduce new control methodologies, algorithms, and systems that address several important issues in developing intelligent, autonomous or semi-autonomous, networked systems for the next generation of UAVs. The papers share a common focus on improved coordination of the members of the networked system to accomplish a common mission, to achieve heightened capability in system reconfiguration to compensate for lost members or connections, and to enhance robustness against terrain complications and attacks.

In their paper, Fariba Fahroo and Michael Ross present the theoretical framework for Pseudospectral (PS) optimal control theory that can be used for design of new maneuvers for UAVs that serve the needs of the mission. Given the expansion of the operational envelope for agile UAVs, design of these maneuvers in an optimal manner has become feasible and the use of advanced numerical techniques for these design problems is essential. In this paper, the authors explore the theoretical (convergence) and implementation issues of these PS controls onboard of UAVs.

Ben G. Fitzpatrick considers certain idempotent modifications of Bayesian analysis which are suitable for decision making in mixed initiative and multi-agent systems in highly autonomous operations. The paper examines Bayesian networks and Bayesian knowledge bases within the context of max-plus probability. Then these techniques are used to treat decentralized control of unmanned air sensing assets.

The paper by Seung Hak Han and William M. McEneaney deals with the problem where one wishes to intercept a vehicle before it reaches a specific target. The vehicle travels on a road network. There are unattended ground sensors (UGSs) on the road network. The data obtained by a UGS is uploaded to an unmanned aerial vehicle (UAV) when the UAV overflies the UGS. There is an interceptor vehicle which may travel the road network. The interceptor uses intruder state estimates based on the data uploaded from the UGSs to the UAV. There is a negative payoff if the intruder is intercepted prior to reaching the target location, and a positive payoff if the intruder reaches the target location without having been previously intercepted. The intruder position is modeled as a Markov chain. The observations are corrupted by random noise. A dynamic programming approach is taken where the value is propagated using a min-plus curse-of-dimensionality-free algorithm. The double-description method for polyhedra is used to reduce computational loads.

Laura R. Humphrey's paper explores the use of model checking for verification in several UAV related applications, including a centralized cooperative control scheme, a decentralized cooperative control scheme, and a scenario involving a human operator. There has been increased interests in verification techniques for complex, autonomous systems. The team might include several autonomous UAVs and unattended ground sensors, as well as a human operator. In such cases, there is a need for techniques to verify the aggregate behavior of the team. Current research suggests that standard software modeling and verification techniques can be extended to meet this need. Depending on whether team coordination is implemented in a centralized or decentralized manner, behavior of a cooperative control system or behavior of the individual agents can be modeled using a formal modeling language, system specifications can be expressed using a modal logic, and a model checker can be used to verify that the modeled behavior of the team meets all specifications. Specifications are written in linear temporal logic. Example models are coded in PROMELA, which are verified using the model checker SPIN.

To determine optimal policies for large scale controlled Markov chains, K. Krishnamoorthy, M. Park, S. Darbha, M. Pachter, and P. Chandler consider a base perimeter patrol stochastic control problem. To determine the optimal control policy, one has to solve a Markov decision problem, whose large size renders exact dynamic programming methods intractable. They propose a state aggregation based approximate linear programming method to construct provably good suboptimal policies instead. The state space is partitioned and the optimal cost-to-go or value function is approximated by a constant over each partition. By minimizing a non-negative cost function defined on the partitions, one can construct an approximate value function that is an upper bound for the optimal value function of the original Markov chain. They show that this approximate value function is independent of the non-negative cost function (or state dependent weights; as it is referred to in the literature) and moreover, this is the least upper bound that one can obtain, given the partitions. Furthermore,we show that the restricted system of linear inequalities also embeds a family of Markov chains of lower dimension, one of which can be used to construct a tight lower bound on the optimal value function. In general, the construction of the lower bound requires the solution to a combinatorial problem.

Khanh Pham's paper considers a class of distributed stochastic systems where interconnected systems closely keep track of reference signals issued by a coordinator. Much of the existing literature concentrates on conducting decisions and control synthesis based solely on expected utilities and averaged performance. However, research in psychology and behavioral decision theory suggests that performance risk plays an important role in shaping preferences in decisions under uncertainty. A new equilibrium concept, called "person-by-person equilibrium" for local best responses, is proposed to analyze signaling effects and mutual influences between an incumbent system, its coordinator and immediate neighbors. Individual member objectives are defined by the multi-attribute utility functions that capture both performance expectation and risk measures to model the satisfaction associated with local best responses with risk-averse attitudes. The problem class and approach of coordination control of distributed stochastic systems proposed here are applicable to and exemplified in military organizations and flexibly autonomous systems.

In the paper of Le Yi Wang and G. Yin, a team of UAVs in surveillance operations aims to achieve fast deployments, robustness against uncertainties and adversaries, and adaptability when the team expands or reduces under time-varying and local communication connections. This paper introduces a new framework for UAV control based on the emerging consensus control for networked systems. Due to unique features of UAV tasks, the consensus control problem becomes weighted and constrained, beyond the typical consensus formulation. Using only neighborhood communications among UAV members, the team coordination achieves a global desired deployment. Algorithms are introduced and their convergence properties are established. It is shown that the algorithms achieve asymptotically the Cramér-Rao lower bound, and hence is optimal among all algorithms in terms of MS errors. Performance optimization within network topology constraints is further explored, including optimal convergence rates and optimal robustness. Examples and case studies demonstrate convergence, robustness, and scalability of the algorithms.

This book is written for researchers, engineers, practitioners, and students in UAV research and related fields. Selected materials from the book may also be used in a graduate seminar course on special topics in UAVs.

Without the help of many individuals, the book project would not have been completed. First, we would like to thank all the authors for their contributions. This book project was initiated during several visits of George Yin and Le Yi Wang to the Air Force Research Laboratory, Wright-Patterson Air Force Base. Our special thanks go to Dr. Siva Banda for providing us with the opportunity and for connecting us with the researchers in the Control Science Center of Excellence, Air Vehicles Directorate. Our thanks also go to the Editor of the Series for providing us with an excellent forum and outlet to present our results. Finally, we thank the Springer professionals for finalizing the book.

Arlington, Virginia — Fariba Fahroo
Detroit, Michigan — Le Yi Wang
Detroit, Michigan — George Yin
February 2013

Contents

Chapter 1
Enhancing the Practical Agility of UAVs via Pseudospectral Optimal Control

Fariba Fahroo and I. Michael Ross

Abstract. UAVs have the capability to perform maneuvers that would otherwise be unrealistic to do in an inhabited craft due to human limitations of both response time as well as comfort or black-out limits. Given this expansion of its operational envelope, a natural question to ask is on the design of new maneuvers for UAVs that serve the needs of the mission. One of these needs is a rapid response time. In this context, we explore the design of a minimum-time velocity reversal maneuver for a fixed-wing UAV. Pseudospectral optimal control theory is used to address this problem. We discuss a wide-range of issues from theory to flight implementation. We show that these issues are interdependent and explore key convergence results that are critical for a successful flight. Results for a MONARC UAV are used to illustrate the close connection between theory and practice.

1.1 Introduction

The agility of UAVs can be characterized in terms of solutions to minimum-time optimal control problems. A flight implementation of solutions to these problems requires careful consideration of the mismatch between the model (used in solving the optimal control problem) and the actual vehicle dynamics as well as considerations of the interactions between the inner-loop and outer loop. When these practical considerations are properly taken into account, the resulting problem is typically nonlinear, constrained in state and control spaces with possible interdependencies between these spaces [1].

Fariba Fahroo
Air Force Office of Scientific Research (AFOSR), Arlington, VA
e-mail: fariba.fahroo@afosr.af.mil

I. Michael Ross
Department of Mechanical and Aerospace Engineering, Naval Postgraduate School, Monterey, CA 93943
e-mail: imross@nps.edu

F. Fahroo et al. (Eds.): *Recent Adv. in Res. on Unmanned Aerial Veh.*, LNCIS 444, pp. 1–30.
DOI: 10.1007/978-3-642-37694-8_1

One possible way to solve such UAV problems is to simplify the control loops and the associated dynamics and use "trim" conditions; however, this approach defeats the entire purpose of enhancing agility as trim is anathema to agility. As a means to exploit the total agility of a UAV, we propose to address such problems head-on using pseudospectral (PS) optimal control.

Pseudospectral optimal control theory, a term coined by Ross [1], is the combination of pseudospectral theory and optimal control theory. PS optimal control is the modern method of choice for onboard flight implementation. In fact, the first flight implementation of PS optimal control was in 2006 when a front page article in *SIAM News* [2] announced that it was successfully used to maneuver the International Space Station, a $100B space asset. Since then, PS optimal control techniques have continued to make headlines in maneuver discovery and rapid flight implementations onboard high-value national systems [3, 4, 5, 6]. In all these applications, the software, DIDO [7], was used to generate and implement the solutions. PS controls, as implemented in DIDO and adopted in OTIS [8], continue to be used for routine operations of ground and flight implementations. This is, in part, due to their independently reproducible performance when integrated with standard simulations [9, 10, 11, 12]. Diverse optimal control problems have been solved by various practitioners; examples include space station attitude control [13, 14], ascent guidance [9], interplanetary solar sail mission design [10], supersonic intercept [11, 12], low-thrust Earth-to-Jupiter rendezvous [12], loitering of unmanned aerial vehicles [15, 16], lunar guidance [17], libration-point stationkeeping [18], momentum-dumping [19], launch vehicle trajectory optimization [20], impulsive orbit transfer optimization [21], inert and electrodynamic tether control [22, 23, 24], aircraft terrain following [25], magnetic attitude control [26], quantum control [27], crane-and-pulley problems [28], autonomous operations for UAVs [29, 30], cooperative control of UAVs [31] and obstacle avoidance for UxVs [32]. Solutions to well over a hundred problems are documented in [33].

The most widely used PS methods are the Legendre [34, 35, 36, 37] and Chebyshev PS methods [38, 39], collectively known as the "big two" methods [1]. This is simply because a rich number of convergence theorems have been proven in the literature [40, 41, 42], and hence one may use these methods with a high degree of confidence in their validity. In addition, in practical applications, one may treat the outcome of the big-two methods in terms of a rigorous application of Pontryagin's Principle and accept or reject solutions based on the optimality conditions [43]. This comfort of guarantees is enunciated as the Covector Mapping Principle [41, 44, 45, 46, 47] (CMP) which provides the foundations for generating explicit Covector Mapping Theorems [36, 37, 41, 42]. The proof of this theorem utilizes specific quadrature formulas [48, 49] that have thus far shown to be valid only for the big-two methods, and hence the popularity of this approach.

In 2005, we proposed [50] a PS method based on the Legendre-Gauss-Radau (LGR) grid to solve optimal control problems. This proposal was motivated by a means to handle the singularity problems that arise as a result of transforming the infinite-time domain to a finite time-domain. Additional studies showed [51] a number of other advantages of the shifted LGR grid, particularly for real-time applications. This generated a natural question: does the LGR grid satisfy CMP-consistency? An apparently simple way to investigate this issue is to apply the CMP to derive an explicit covector map [46] in a manner similar to that of the Legendre-Gauss-Lobatto (LGL) grid. While conceptually simple, this task is not altogether straightforward as a key lemma related to an integration-by-parts formula [37] is not readily available for non-LGL grids. This crucial formula identifies the correct finite-dimensional inner-product space that is necessary for the construction of the discretized 1-form that defines the sequences of discretized Lagrangians that converge to the continuous-time Lagrangian [52]. In [53], we showed that a special weight function met all the appropriate criteria for the CMP leading to a new theorem on dual consistency for the LGR grid.

Motivated by these ideas for the LGR grid and its possible connections to the LGL grid, we developed a unified approach to investigate all PS methods in [54, 55]. *A key idea that emerged from the unification principles is the notion of weighed interpolants, their duals, and their direct effect on the generation of the correct pre-Hilbert space where the computed solutions lie.* The concept of a pre-Hilbert space is crucial for proofs of convergence theorems [40, 41, 42] that rely on the separability of the infinite-dimensional Hilbert spaces used to construct highly-accurate solutions to practical optimal control problems. We presented these ideas in a series of papers [33, 51, 53, 54]; here, we expand and clarify our previous results and illustrate its connections to a successful flight implementation.

We begin this paper at the level of first principles by providing a unified set of foundations for all PS methods for optimal control; i.e. PS methods over an arbitrary grid. In limiting the scope of the discussions, we focus on three Legendre grids: LGL, LGR and Legendre-Gauss (LG). From the perspective of unification, it is shown that the LGL grid is indeed the correct Legendre-based PS method for solving boundary value problems (BVPs) with arbitrary boundary conditions and hence all generic finite-horizon optimal control problems. The unification also illustrates why the LGR grid is preferable over the LGL grid for stabilizing control systems as this problem falls under the realm of a special optimal control problem, the specialty being the infiniteness of the horizon [53]. We illustrate some key consequences of our theory by demonstrating how disastrous results are possible when the LGR or LG grids are artificially forced to solve BVP-type problems over a finite horizon.

1.2 A Distilled Optimal Control Problem

For simplicity in exposition, we will consider the following scalar Bolza problem, with the understanding that our discretization methods can be easily extended to higher-dimensional cases. For the same reason we will ignore path constraints as these can also be easily incorporated into our framework by replacing the control Hamiltonian by the Lagrangian of the Hamiltonian [37]. In simplifying such bookkeeping issues, we consider the problem of finding the optimal state-control function pair $t \mapsto (x,u) \in \mathbb{R} \times \mathbb{R}$ that solves the following problem (Problem B):

$$x \in \mathbb{X} \subseteq \mathbb{R}, \quad u \in \mathbb{R}$$

$$(B)\begin{cases} \text{Minimize} & J[x(\cdot),u(\cdot)] = E(x(-1),x(1)) + \int_{-1}^{1} F(x(t),u(t))\,dt \\ \text{Subject to} & \dot{x}(t) = f(x(t),u(t)) \\ & e(x(-1),x(1)) = 0 \end{cases}$$

where $\mathbb{X}$ is an open set in $\mathbb{R}$ and the problem data (i.e. the endpoint cost function, E, the running cost function, F, the vector field, f and the endpoint constraint function, e) are assumed to be at least C^1-smooth. An application of Pontryagin's Minimum Principle (PMP) results in a two-point BVP that we denote as Problem B^λ. This "dualization" is achieved through the construction of the control Hamiltonian, H, and endpoint Lagrangian, $\bar{E}$, defined as

$$H(\lambda,x,u) = F(x,u) + \lambda f(x,u) \tag{1.1}$$

$$\bar{E}(\nu,x(-1),x(1)) = E(x(-1),x(1)) + \nu e(x(-1),x(1)) \tag{1.2}$$

where λ is the adjoint covector (or costate) and ν is the endpoint covector. Problem B^λ is now posed as follows:

$$(B^\lambda)\begin{cases} \text{Find} & t \mapsto (x,u,\lambda) \text{ and } \nu \\ \text{Such that} & \dot{x}(t) - f(x(t),u(t)) = 0 \\ & e(x(-1),x(1)) = 0 \\ & \dot{\lambda}(t) + \partial_x H(\lambda(t),x(t),u(t)) = 0 \\ & \partial_u H(\lambda(t),x(t),u(t)) = 0 \\ & \lambda(-1) + \dfrac{\partial \bar{E}}{\partial x(-1)}(\nu,x(-1),x(1)) = 0 \\ & \lambda(1) - \dfrac{\partial \bar{E}}{\partial x(1)}(\nu,x(-1),x(1)) = 0 \end{cases}$$

In using the PMP to solve a trajectory optimization problem, we must solve Problem B^λ which is clearly a problem of finding the zeros of a map in an appropriate function space [1]. This is the so-called indirect method. From a first-principles perspective [43], the PMP is used as follows: For every optimal solution to Problem B, there exist an adjoint covector function, $t \mapsto \lambda$, and an endpoint covector, ν, that satisfy the conditions set forth by Problem B^λ. Thus, every candidate optimal solution to Problem B must also be a solution to Problem B^λ under appropriate technical conditions. The LGL grid ensures a satisfaction of this condition through its Covector Mapping Theorem [36, 37]. We present a generalized and unified version of this theorem with respect PS grid points. A generalized version of this theorem with respect to the stability and convergence of the approximation is described by Gong et al [41].

1.3 A Unified Perspective on Pseudospectral Optimal Control

Legendre PS methods are a subset of a larger class of spectral methods which use orthogonal basis functions in global expansions similar to Fourier and Sinc series expansions [56]. What distinguishes PS methods is that they are based on approximating the unknown functions by interpolants [57]. This is one reason why PS methods are sharply different from other polynomial methods previously considered in the literature. The interpolating points in a PS method can be any of the three major classes of Gaussian grid points called nodes: Gauss points, Radau points and Lobatto points. Note that Radau and Lobatto points are also Gaussian quadrature points. All of these nodes are zeros of the orthogonal polynomials or their derivatives such as Legendre and Chebyshev or more generally the Jacobi polynomials. Thus, a very large family of PS methods can be generated to solve optimal control problems. This notion is similar to Runge-Kutta (RK) methods. For example, Euler, trapezoid, Hermite-Simpson and many other methods are equivalently some form of an RK method with appropriately chosen coefficients [58]. Just as not all RK methods are legitimate, not all PS methods are legitimate. This is particularly true in solving optimal control problems. For example, an RK method that is legitimate for propagating an ODE may fail gloriously when applied to an optimal control problem. In other words, it is imperative to carefully select an appropriate method to solve optimal control problems.

In the absence of special information (such as special boundary conditions), it is customary to select Chebyshev PS methods in solving PDE problems as they are simple and effective [57]. For optimal control of ODEs, the Legendre PS method is preferable as it maintains the consistency of dual approximations. In 2005, we proposed the usage of Radau points [50, 51, 53] as a means to manage a singularity problem that arises in solving infinite-horizon control problems. Subsequently, we expanded the scope to present a more unified view of these ideas to show the similarities and differences of the methods based on the choice of these different nodes and show the effect of these nodes on the discretization of both the primal and dual problems [54].

1.3.1 Weighted Interpolants and Differentiation Matrices on Arbitrary Grids

We begin with the basic definition of PS methods and the different ingredients required in defining these methods for various nodes. Following [54], we select PS methods based on weighted interpolants of the form,

$$y(t) \approx y^N(t) = \sum_{j=0}^{N} \frac{W(t)}{W(t_j)} \phi_j(t) y_j, \; a \le t \le b \tag{1.3}$$

where $y(t)$ is an arbitrary function. Here the nodes $t_j, j = 0,...,N$ are a set of distinct interpolation nodes (defined below) in the interval $[a,b]$, the weight function $W(t)$ is a positive function on the interval, and $\phi_j(t)$ is the $Nth-$ order Lagrange interpolating polynomial that satisfies the relationship $\phi_j(t_k) = \delta_{jk}$. This implies that

$$y_j = y^N(t_j), \ \ j = 0,...N. \tag{1.4}$$

An expression for the Lagrange polynomial can be written as [56]

$$\phi_j(t) = \frac{g_N(t)}{g'_N(t_j)(t-t_j)}, \ g_N(t) = \prod_{j=0}^{N}(t-t_j). \tag{1.5}$$

One important tenant of polynomial approximation of functions is that differentiation of the approximated functions can be performed by differentiation of the interpolating polynomial,

$$\frac{dy^N(t)}{dt} = \sum_{j=0}^{N} \frac{y_j}{W(t_j)}[W'(t)\phi_j(t) + W(t)\phi'_j]$$

Since only the values of the derivative at the nodes t_i are required for PS methods, then we have,

$$\left.\frac{dy^N(t)}{dt}\right|_{t_i} = \sum_{j=0}^{N} \frac{y_j}{W(t_j)}[W'(t_i)\delta_{ij} + W(t_i)D_{ij}] = \sum_{j=0}^{N} D_{ij}[W]y_j \tag{1.6}$$

where we use $D_{ij}[W]$ as a shorthand notation for the W-weighted differentiation matrix,

$$D_{ij}[W] := \frac{[W'(t_i)\delta_{ij} + W(t_i)D_{ij}]}{W(t_j)} \tag{1.7}$$

and D_{ij} is usual unweighted differentiation matrix given by,

$$D_{ij} := \left.\frac{d\phi_j(t)}{dt}\right|_{t=t_i} \tag{1.8}$$

Thus, when $W(t) = 1$, we have

$$D_{ij}[1] = D_{ij} \tag{1.9}$$

From Eq. (1.5), the unweighted differentiation matrix, $D_{ij} = \phi'_j(t_i)$, has the form,

$$D_{ij} = \begin{cases} \dfrac{g'_N(t_i)}{g'_N(t_j)} \dfrac{1}{(t_i - t_j)}, & i \neq j \\ \\ \dfrac{g''_N(t_i)}{2g'_N(t_i)}, & i = j \end{cases} \tag{1.10}$$

The above equations are the general representations of the derivative of the Lagrange polynomials evaluated at arbitrary interpolation nodes. Thanks to Runge, it is well-known that an improper selection of the grid points can lead to disastrous consequences. In fact, a uniform distribution of grid points is the worst possible choice for polynomial interpolation and hence differentiation. On the other hand, the best possible choice of grid points for integration, differentiation and interpolation of functions are Gaussian quadrature points [48]. Consequently, all PS methods use Gaussian quadrature points. In order to limit the scope of this paper, we limit our discussions to Gaussian points based on Legendre polynomials as these polynomials offer elegant theoretical properties for optimal control applications.

1.3.2 Legendre Nodes

Gaussian quadrature points lie in the interval $[-1,1]$. In most problems the physical problem is posed on some interval $[a,b]$ which can then be easily transformed to the computational domain $[-1,1]$ via a linear transformation.

The Gauss quadrature points are zeros that are interior to the interval $[-1,1]$. The Gauss-Radau zeros include one of the end points of the interval, usually the left-end point at $t=-1$. The Gauss-Lobatto points include both endpoints of the interval at $t=-1$, and $t=1$; see Fig. 1.1. These quadrature nodes are related to the zeros of

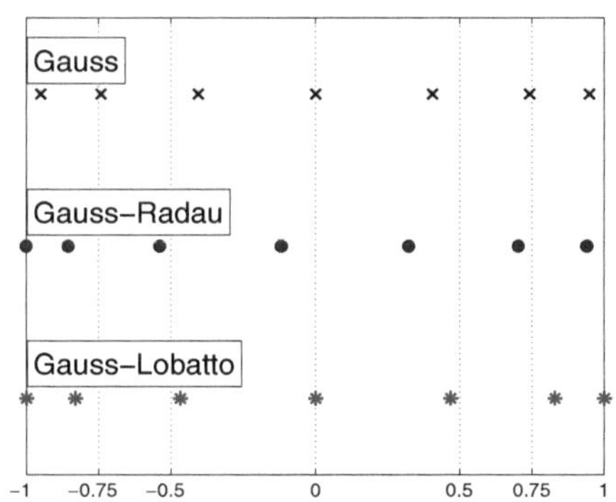

Fig. 1.1 Illustrating the various Legendre quadrature nodes: Legendre-Gauss (LG), Legendre-Gauss-Radau (LGR) and Legendre-Gauss-Lobatto (LGL) grid points

the Qth-order Jacobi polynomials $P_Q^{\alpha,\beta}$ which are orthogonal on the interval $[-1,1]$ with respect to the inner product [48, 49],

$$\int_{-1}^{1}(1-t)^{\alpha}(1+t)^{\beta}P_Q^{\alpha,\beta}(t)P_{Q'}^{\alpha,\beta}(t)dt \tag{1.11}$$

Note that the Legendre polynomials are a special case of Jacobi polynomials which correspond to $\alpha=\beta=0$. Let $t_{i,Q}^{\alpha,\beta}$ be the Q zeros of the Qth-order Jacobi polynomial $P_Q^{\alpha,\beta}$ such that

$$P_Q^{\alpha,\beta}(t_{i,Q}^{\alpha,\beta})=0, \quad i=0,...,Q-1 \tag{1.12}$$

Then, we can define zeros and weights that approximate the following integral

$$\int_{-1}^{1} u(t)dt=\sum_{i=0}^{N} w_i u(t_i)+R(u) \tag{1.13}$$

and find expressions for the unweighted differentiation matrices for these nodes as follows [49]:

1.3.2.1 Legendre-Gauss (LG)

$$t_i=t_{i,N+1}^{0,0}, \quad i=0,\ldots,N$$

$$w_i^{0,0}=\frac{2}{1-(t_i)^2}\left[\frac{d}{dt}(L_{N+1}(t)|_{t=t_i}\right]^{-2}, \quad i=0,\ldots,N,$$

$$R(u)=0 \text{ if } u(t)\in\mathscr{P}^{2N+1}([-1,1]).$$

$$D_{ij}=\begin{cases}\frac{L'_{N+1}(t_i)}{L'_{N+1}(t_j)(t_i-t_j)}, & i\neq j, 0\leq i,j\leq N\\ \frac{t_i}{1-t_i{}^2}, & i=j\end{cases}$$

1.3.2.2 Legendre-Gauss-Radau (LGR)

$$t_i=\begin{cases}-1, & i=0,\\ t_{i-1,N}^{0,1}, & i=0,\ldots,N\end{cases}$$

$$w_i^{0,0}=\frac{1-t_i}{(N+1)^2[L_N(t_i)]^2}, \; i=0,\ldots,N,$$

$$R(u)=0 \text{ if } u(t)\in\mathscr{P}^{2N}([-1,1]).$$

$$D_{ij}=\begin{cases}\frac{-N(N+2)}{4}, & i=j=0,\\ \frac{L_N(t_i)}{L_N(t_j)}\frac{1-t_j}{1-t_i}\frac{1}{(t_i-t_j)}, & i\neq j, 0\leq i,j\leq N\\ \frac{1}{2(1-t_i)}, & 1\leq i=j\leq N\end{cases}$$

1.3.2.3 Legendre-Gauss-Lobatto (LGL)

$$t_i = \begin{cases} -1, & i=0, \\ t^{1,1}_{i-1,N-1}, & i=1,\dots,N-1 \\ 1 & i=1 \end{cases}$$

$$w^{0,0}_i = \frac{2}{N(N+1)[L_N(t_i)]^2}, i=0,\dots,N,$$

$$R(u)=0 \text{ if } u(t)\in \mathscr{P}^{2N-1}([-1,1]).$$

$$D_{ij} = \begin{cases} \frac{-N(N+1)}{4}, & i=j=0, \\ \frac{L_N(t_i)}{L_N(t_j)}\frac{1}{(t_i-t_j)}, & i\neq j, 0\leq i,j\leq N \\ 0, & 1\leq i=j\leq N-1 \\ \frac{N(N+1)}{4}, & i=j=N, \end{cases}$$

where $\mathscr{P}^N$ is the set of all algebraic polynomials of degree at most N.

1.3.3 Pre-Hilbert Spaces

Given the definitions of the quadrature nodes and weights above, we can proceed with the definitions of the appropriate discrete inner-product spaces associated with the above nodes and weights. Let $[-1,1]\mapsto \{y(t),z(t)\}$ be real-valued functions in $L^2([-1,1],\mathbb{R})$. The *standard* inner product in L^2 is given by

$$\langle y,z\rangle_{L^2} := \int_{-1}^{+1} y(t)z(t)\,dt$$

For any given $m\in\mathbb{N}$, and $\{w_j>0,\ j=0,1,\dots m\}$, define the *weighted discrete* inner product in L^2 to be

$$\langle y,z\rangle_{m,w} := \sum_{j=0}^{m} y(t_j)w_j z(t_j)$$

For the various Gauss quadrature nodes and the standard inner-product in L^2, we have the following result [56]:

Lemma 1. *For all* $pq\in\mathscr{P}^{2N+\zeta}$,

$$\langle p,q\rangle_{L^2} = \langle p,q\rangle_{N,w}$$

where $\zeta=1$ *for Legendre-Gauss (LG),* $\zeta=0$ *Legendre-Gauss-Radau (LGR) and* $\zeta=-1$ *for Legendre-Gauss-Lobatto (LGL) integration and weights.*

For any weight function, $W(t)$, the *weighted* inner product in L^2 is given by

$$\langle y,z\rangle_{L^2_W} := \int_{-1}^{+1} y(t)W(t)z(t)\,dt$$

Let $W(t)$ be any one of the following weight functions,

$$W(t) = \begin{cases} W_{lgl}(t) & \Rightarrow W(t) = 1 \\ W_{lgr}(t) & \Rightarrow W(t) = 1-t \\ W_{lg}(t) & \Rightarrow W(t) = 1-t^2 \end{cases} \tag{1.14}$$

From Lemma 1, we have the following unified result:

Lemma 2. *For all $pq \in \mathscr{P}^{2N-1}$,*

$$\langle p,q\rangle_{L^2_W} = \langle p,q\rangle_{N,w}$$

Remark 1. Lemma 2 implies that W_{lgl} is the only weight function that is dual to itself in the standard inner product space and unweighted interpolation.

This primal-dual symmetry explains why the LGL grid offers elegant properties.

Lemma 3. *For the three Legendre quadrature nodes, LG, LGR and LGL, the following relationships hold: For $i \neq j$ we have,*

$$D_{ij} = -\frac{W(t_j)}{W(t_i)}\frac{w_j}{w_i}D_{ji}$$

Hence, we can write,

$$D_{ij}[W] = -\frac{w_j}{w_i}D_{ji},\; i \neq j$$

and for $i = j$ the following relationships hold:

$$\textit{For LG Nodes} \rightarrow D_{ii}[W_{lg}] = -D_{ii},\, i = 0,\cdots,N \tag{1.15}$$

$$\textit{For LGR Nodes} \rightarrow \begin{cases} D_{ii}[W_{lgr}] = -D_{ii}\; i = 1,\cdots,N \\ D_{00}[W_{lgr}] = -D_{00} - \frac{1}{w_0} \end{cases} \tag{1.16}$$

$$\textit{For LGL Nodes} \rightarrow \begin{cases} D_{ii}[W_{lgl}] = D_{ii} = 0,\;\; i = 1,\cdots,N-1 \\ D_{00}[W_{lgl}] = -D_{00} - 1/w_0 \\ D_{NN}[W_{lgl}] = -D_{NN} + 1/w_N \end{cases} \tag{1.17}$$

1.4 Primal-Dual Pseudospectral Discretizations

We use the following expansions [54] for approximating the state and costate variables,

$$x^N(t) = \sum_{j=0}^{N} \frac{W(t)}{W(t_j)} x_j \phi_j(t), \quad \lambda^N(t) = \sum_{j=0}^{N} \frac{W^*(t)}{W^*(t_j)} \lambda_j \phi_j(t) \tag{1.18}$$

where W and W^* are appropriate choices of weight functions. Lemma 2 suggests that if we take $W(t) = 1$ and LGL nodes, then we must take $W^*(t) = 1$ as well. This is in conformance with the standard theory [37]. It is clear that the use of Lemma 2 in the integration-by-parts argument proposed in [35, 37], generalizes the standard theory by suggesting that if we take $W(t) = 1$ and LGR nodes, then $W^*(t) = 1 - t$; likewise, for $W(t) = 1$ and LG nodes, $W^*(t) = 1 - t^2$. Similarly, if we take $W(t) = 1 - t$ for LGR nodes, then $W^*(t) = 1$ etc.; thus,

$$(W^*(t))^* = W(t) \tag{1.19}$$

That is, the dual of the dual is the original function.

For approximating the control variables, we take

$$u^N(t) = \sum_{j=0}^{N} u_j \psi_j(t) \tag{1.20}$$

where $\psi_j(t)$ is any interpolant. *Note this key point that constitutes part of the unification principles* [40, 41]. That is $u^N(t)$ is not necessarily constructed as a Lagrange interpolating polynomial; rather, $u^N(t)$ is constructed using any type of interpolation such as linear, cubic, spline etc. In previous versions, it was assumed that $u^N(t)$ must be Lagrange interpolants leading to certain convergence questions with increase in N. By choosing $u^N(t)$ to be any interpolant, we retain the key elements of pseudospectral theory while allowing $\psi_j(t)$ to be chosen arbitrarily. This arbitrariness can then be exploited to further enhance the spectral convergence property of PS methods by collecting information in between nodes through an application of Bellman's Principle. Thus ψ_j becomes an abstract interpolant that may be selected based on optimization principles. These ideas have been fully discussed by Ross et al [59, 60] in formulating the Bellman pseudospectral method.

The preceding points illustrate how PS methods for optimal control have indeed evolved to a separate and distinct theory from those used to solve problems in fluid dynamics, the original source of ideas for PS methods for optimal control.

1.4.1 Discretization of the Primal Problem (B^N)

Based on these unified ideas, we formulate PS methods for optimal control as follows: From Eqs. (1.6) and (1.18), it follows that the state derivative is approximated as,

$$\left.\frac{dx^N(t)}{dt}\right|_{t_i} = \sum_{j=0}^{N} D_{ij}[W] x_j \tag{1.21}$$

Suppose we set $W(t) \equiv 1$ for the state variables for all Legendre PS methods. This automatically defines dual weight functions for costate interpolation. We may also set $W(t)$ according to the appropriate class of weighted interpolations in which case we would arrive at the corresponding duals. Choosing $W(t) \equiv 1$ for the primal variables implies that the primal differentiation matrix, $D[W]$, is the standard differentiation matrix (Cf. Eq. (1.9)). Applying the quadrature rules for approximating the integral terms and using the appropriate form of the derivative matrix according to the choice of the nodes, we have the following expressions for Problem B^N which is the discretized form of Problem B:

$$X \in \mathbb{R}^{N_n}, \quad U \in \mathbb{R}^{N_n}$$

$$(B^N) \begin{cases} \text{Minimize} & J^N[X,U] = E(x_0,x_N) + \sum_{j=0}^{N} w_j F(x_j,u_j) \\ \text{Subject to} & \sum_{j=0}^{N} D_{ij}x_j - f(x_i,u_i) = 0 \quad i = 0,1,\ldots,N \\ & e(x_0,x_N) = 0 \end{cases}$$

where $X = [x_0, x_1, \ldots, x_{N_n}]$ and $U = [u_0, u_1, \ldots, u_{N_n}]$.

1.4.2 *Selection of a Proper Choice of PS Grids*

Lemma 2 suggests the proper choice of an appropriate PS method for Problem B^N. In many optimal control problems the endpoints are not totally free and are frequently constrained; that is, the endpoint function, e, is typically a function of both $x(-1)$ and $x(1)$. Then, according to Lemma 2, we must choose the LGL grid. Now suppose that we choose an LG or LGR grid for such a problem. The motivation for this selection may be based on Lemma 1 which suggests the apparently higher degree of approximation for LG and LGR nodes when compared to LGL nodes. That this is not true is best illustrated by a simple counter example of Gong et al [40]:

$$\mathbf{x} = (x,v) \in \mathbb{R}^2, \quad u \in \mathbb{R}$$

$$(P_1) \begin{cases} \text{Minimize} & J[\mathbf{x}(\cdot),u(\cdot)] = \int_0^1 v(t)u(t)\,dt \\ \text{Subject to} & \dot{x}(t) = v(t) \\ & \dot{v}(t) = -v(t) + u(t) \\ & v(t) \geq 0 \\ & 0 \leq u(t) \leq 2 \\ & (x(0),v(0)) = (0,1) \\ & (x(1),v(1)) = (1,1) \end{cases}$$

This problem describes a particle of unit mass ($m = 1$) moving in a linear resistive medium ($\alpha = -1$) where x is the position, v is the velocity and u is the applied force; see Fig. 1.2. The optimal control problem is to minimize the total amount of work done. From physical considerations or by a direct application of the Minimum Principle, it is easy to verify that the optimal control is a constant and is equal to the amount of force required to maintain the initial speed. Thus, the exact optimal control is given by,

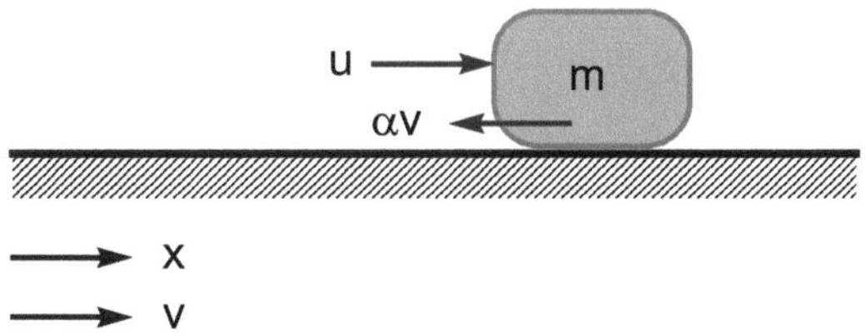

Fig. 1.2 Schematic for Problem P_1

$$u^*(t) = 1 \tag{1.22}$$

Since the optimal control is a polynomial, a correct PS method is expected to achieve exact performance for sufficiently large N. The solution to this problem based on the standard LGL discretization is shown in Fig. 1.3 for 10 nodes. Predictably, the LGL

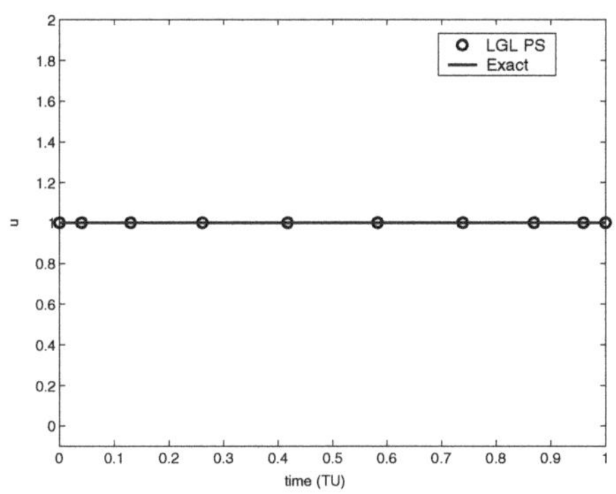

Fig. 1.3 Control solution to Problem P_1 over an LGL grid of $N = 10$

grid generates the exact solution. On the other hand, the LG and LGR grids produce disastrous solutions as shown in Fig. 1.4. Thus, choosing a PS grid for optimal control based on Lemma 1 alone is a bad proposition. To illustrate the point that the LG and LGR grids are not just converging at a slower rate than the LGL grid, their control solutions are plotted in Fig. 1.5 for $N = 30$. It is clear that the solutions over LG and LGR grids do not converge. What is most interesting about the lack of convergence of these solutions is that the grid points converge to the LGL grid points as N increases; see Fig. 1.6.

Thus, a convergence of the grid points is, obviously, not a sufficient condition for the convergence of the solutions.

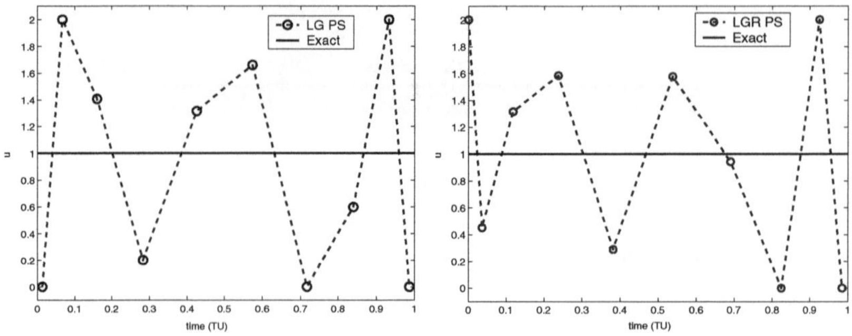

Fig. 1.4 Control solutions to Problem P_1 for LGR and LG grids of $N = 10$; compare Fig. 1.3

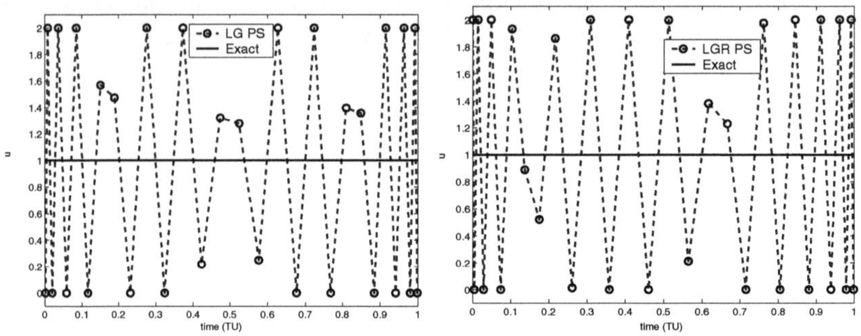

Fig. 1.5 Control solutions to Problem P_1 for $N = 30$: Illustrating why the LGR and LG grids are the wrong choices for this problem

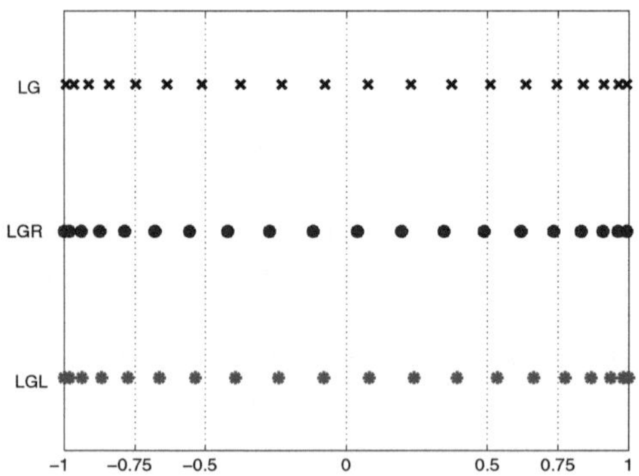

Fig. 1.6 Illustrating the near absence of differences between the three Legendre grids for increasing N

1.4.3 Discretization of the Dualized Problem ($B^{\lambda N}$)

Given that dual space considerations are indeed important even when only primal variables are sought [46], we now explore the discretization of the dualized problem.

From Eqs. (1.7) and (1.18), it follows that the costate derivative is approximated as,

$$\left.\frac{d\lambda^N(t)}{dt}\right|_{t_i} = \sum_{j=0}^{N} D^*_{ij}\lambda_j \tag{1.23}$$

with

$$D^*_{ij} = D_{ij}[W] = \frac{[W'(t_i)\delta_{ij} + W(t_i)D_{ij}]}{W(t_j)}$$

Thus Problem $B^{\lambda N}$ which is the discretization of Problem B^λ defined in Sec. II, can be constructed as,

$$(B^{\lambda N}) \begin{cases} \sum_{j=0}^{N} D_{ij}x_j - f(x_i,u_i) = 0 & i = 0,1,\ldots,N \\ e(x_0,x_N) = 0 & \\ \sum_{j=0}^{N} D^*_{ij}\lambda_j + \partial_{x_i}H(\lambda_i,x_i,u_i) = 0 & i = 0,1,\ldots,N \\ \partial_{u_i}H(\lambda_i,x_i,u_i) = 0 & i = 0,1,\ldots,N \\ \lambda_0 + \frac{\partial \bar{E}}{\partial x_0}(\nu,x_0,x_N) = 0 & \\ \lambda_N - \frac{\partial \bar{E}}{\partial x_N}(\nu,x_0,x_N) = 0 & \end{cases}$$

Thus, in the unified PS method, we use the matrices D and D^* as primal and dual differentiation matrices arising from the primal and dual weighted interpolation with weight functions, $W(t)$ and $W^*(t)$, respectively. Furthermore, from Eq. (1.19), it follows that we may choose $W(t)$ for discretizing the dual, in which case we need to use $W^*(t)$ for discretizing the primals.

1.5 A Unified Covector Mapping Theorem

Based on the preceding discussions it is now abundantly clear that the correct inner product space for all three PS discretizations is given by $\mathbb{R}^{N_n}_W$, the finite-dimensional Hilbert space $\mathbb{R}^{N_n}$ equipped with the *weighted* inner product,

$$\langle a,b\rangle_{\mathbb{R}^{N_n}_W} := \sum_{i=0}^{N} a_i w_i b_i \qquad a,b \in \mathbb{R}^{N_n}$$

Thus, the Lagrangian in $\mathbb{R}^{N_n}_W$ is given by [52],

$$\overline{J^N}[\widetilde{\lambda},\widetilde{\nu},x,u] = \sum_{j=0}^{N} w_j F(x_j,u_j) + \sum_{i=0}^{N} w_i \widetilde{\lambda}_i f(x_i,u_i) - \sum_{i=0}^{N} w_i \widetilde{\lambda}_i \sum_{j=0}^{N} D_{ij} x_j + \widetilde{\nu} e(x_0,x_N) + E(x_0,x_N)$$

$$= \sum_{i=0}^{N} w_i H(\widetilde{\lambda}_i,x_i,u_i) - \sum_{i=0}^{N} w_i \widetilde{\lambda}_i \sum_{j=0}^{N} D_{ij} x_j + \bar{E}(\widetilde{\nu},x_0,x_N)$$

where $\widetilde{\lambda}$ and $\widetilde{\nu}$ are the Karush-Kuhn-Tucker (KKT) multipliers in $\mathbb{R}_w^{N_n}$. From the KKT theorem we have,

$$\partial_{x_k} \overline{J^N}[\widetilde{\lambda},\widetilde{\nu},x,u] = w_k \partial_{x_k} H(\widetilde{\lambda}_k,x_k,u_k) - \sum_{i=0}^{N} w_i \widetilde{\lambda}_i \sum_{j=0}^{N} D_{ij} \delta_{jk} \qquad k = 1,\ldots,N-1$$

$$= w_k \partial_{x_k} H(\widetilde{\lambda}_k,x_k,u_k) - \sum_{i=0}^{N} w_i \widetilde{\lambda}_i D_{ik}$$

From Lemma 3 we have,

$$D^*_{ki} := -\frac{w_i}{w_k} D_{ik} \quad for\ k = 1,\ldots,N-1, i = 0,\ldots,N \tag{1.24}$$

Hence,

$$\partial_{x_k} \overline{J^N}[\widetilde{\lambda},\widetilde{\nu},X,U] = w_k \left(\partial_{x_k} H(\widetilde{\lambda}_k,x_k,u_k) + \sum_{i=0}^{N} D^*_{ki} \widetilde{\lambda}_i \right) = 0$$

Similarly,

$$\partial_{x_k} \overline{J^N}[\widetilde{\lambda},\widetilde{\nu},X,U] = w_k \partial_{x_k} H(\widetilde{\lambda}_k,x_k,u_k) - \sum_{i=0}^{N} w_i \widetilde{\lambda}_i \sum_{j=0}^{N} D_{ij} \delta_{jk} + \widetilde{\nu} \partial_{x_k} e(x_0,x_N), \quad k = 0,N$$

$$= w_k \partial_{x_k} H(\widetilde{\lambda}_k,x_k,u_k) - \sum_{i=0}^{N} w_i \widetilde{\lambda}_i D_{ik} + \widetilde{\nu} \partial_{x_k} e(x_0,x_N)$$

From Lemma 3, this implies,

$$\partial_{x_k} \overline{J^N}[\widetilde{\lambda},\widetilde{\nu},X,U] = w_k \left(\partial_{x_k} H(\widetilde{\lambda}_k,x_k,u_k) + \sum_{i=0}^{N} D^*_{ki} \widetilde{\lambda}_i \right) + \widetilde{\lambda}^*_k + \widetilde{\nu} \partial_{x_k} e(x_0,x_N) = 0, \quad k = 0$$

$$\partial_{x_k} \overline{J^N}[\widetilde{\lambda},\widetilde{\nu},X,U] = w_k \left(\partial_{x_k} H(\widetilde{\lambda}_k,x_k,u_k) + \sum_{i=0}^{N} D^*_{ki} \widetilde{\lambda}_i \right) + \widetilde{\lambda}^*_k - \widetilde{\nu} \partial_{x_k} e(x_0,x_N) = 0, \quad k = N$$

where $\widetilde{\lambda}_k^*$ is given by:

1.5.0.1 Legendre-Gauss

$\widetilde{\lambda}_k^* = 0$ for $k = 0$ and N

1.5.0.2 Legendre-Gauss-Radau

$\widetilde{\lambda}_k^* = \widetilde{\lambda}_k$ for $k = 0$ and $\widetilde{\lambda}_k^* = 0$ for $k = N$

1.5.0.3 Legendre-Gauss-Lobatto

$\widetilde{\lambda}_k^* = \widetilde{\lambda}_k$ for $k = 0$ and N

Thus Problem $B^{N\lambda}$ can be written as,

$$
(B^{N\lambda}) \begin{cases}
\sum_{j=0}^{N} D_{ij} x_j - f(x_i, u_i) = 0 & i = 0, 1, \ldots, N \\
e(x_0, x_N) = 0 & \\
\sum_{j=0}^{N} D_{ij}^* \widetilde{\lambda}_j + \partial_{x_i} H(\widetilde{\lambda}_i, x_i, u_i) = 0 & i = 1, \ldots, N-1 \\
\partial_{u_i} H(\widetilde{\lambda}_i, x_i, u_i) = 0 & i = 0, 1, \ldots, N \\
\sum_{j=0}^{N} D_{ij}^* \widetilde{\lambda}_j + \partial_{x_i} H(\widetilde{\lambda}_i, x_i, u_i) = -c_0^* & i = 0 \\
\sum_{j=0}^{N} D_{ij}^* \widetilde{\lambda}_j + \partial_{x_i} H(\widetilde{\lambda}_i, x_i, u_i) = -c_N^* & i = N \\
\widetilde{\lambda}_0^* + \frac{\partial \bar{E}}{\partial x_0}(\widetilde{\nu}, x_0, x_N) = w_0 c_0^* & \\
\widetilde{\lambda}_N^* - \frac{\partial \bar{E}}{\partial x_N}(\widetilde{\nu}, x_0, x_N) = w_N c_N^* &
\end{cases}
$$

where c_0^* and c_N^* are arbitrary real numbers.

Now suppose that a solution to Problem $B^{\lambda N}$ exists. That is, we suppose that a (discretized) solution to the two-point BVP exists. Then the covectors of Problem $B^{\lambda N}$ provide an existence theorem for the solution of Problem $B^{N\lambda}$; that is, for all $k = 0, \ldots, N$, $\widetilde{\lambda}_k = \lambda_k$ and $\widetilde{\nu} = \nu$ under the following conditions, a solution to Problem $B^{N\lambda}$ exists:

Condition A:

Legendre-Gauss

x_0 and x_N are free; i.e. unspecified.

Legendre-Gauss-Radau

x_0 may be arbitrarily specified but x_N must be free, or vice versa.

Legendre-Gauss-Lobatto

Both x_0 and x_N may be arbitrarily specified.

The dual to Condition A is given by,

Condition A^*:

Legendre-Gauss

$x_0 = 0 = x_N$.

Legendre-Gauss-Radau

x_0 may be arbitrarily specified but $x_N = 0$, or vice versa.

Legendre-Gauss-Lobatto

Both x_0 and x_N may be arbitrarily specified.

Thus, any solution to Problem $B^{\lambda N}$ is also a solution to Problem $B^{N\lambda}$; in this case, we have,

$$c_0^* = 0 = c_N^* \tag{1.25}$$

That is, Eq. (1.25) is a matching condition that matches Problem $B^{N\lambda}$ to Problem $B^{\lambda N}$. Matching conditions [61, 62] are part of the totality of closure conditions required to complete the circuit (arrows) indicated in Fig. 1.7.

Fig. 1.7 is a commutative diagram promulgated in [44] that forms part of the broader notion of the Covector Mapping Principle in optimal control. A historical account of the origins of this diagram and the ideas embedded in it are documented in [45]. *The concepts embedded in Fig. 1.7 form yet another unification principle in PS methods and optimal control at large.*

1.5.1 A Unified Pseudospectral Covector Mapping Theorem

In completing the steps suggested in Fig. 1.7 towards the development of a Unified Covector Mapping Theorem, we identify the following multiplier sets analogous to those introduced in [63]: Let $\chi := \{[\mathbf{x}_k], [\mathbf{u}_k]\}$ and $\Lambda := \{\nu_0, [\mu_k], [\lambda_k]\}$. We denote by $\mathbb{M}^{\lambda N}(\chi)$ the multiplier set corresponding to χ,

$$\mathbb{M}^{\lambda N}(\chi) := \left\{\Lambda : \Lambda \text{ satisfies conditions of Problem } B^{\lambda N}\right\} \tag{1.26}$$

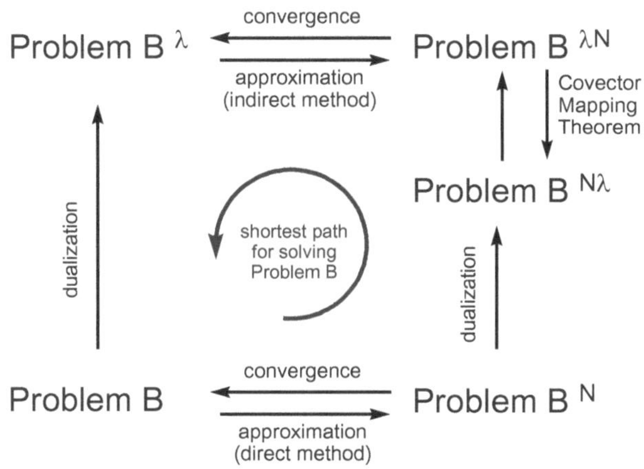

Fig. 1.7 Illustrating the Covector Mapping Principle and the unification of direct and indirect methods in optimal control as first described by Ross and Fahroo [36, 37, 44]

Similarly, we define, $\widetilde{\Lambda} := \left\{\widetilde{\nu}_0, [\widetilde{\mu}_k], [\widetilde{\lambda}_k]\right\}$ and $M^{N\lambda}(\chi)$ the multiplier set,

$$\mathbb{M}^{N\lambda}(\chi) := \left\{\widetilde{\Lambda} : \widetilde{\Lambda} \text{ satisfies conditions of Problem } B^{N\lambda}\right\} \tag{1.27}$$

Clearly, $\mathbb{M}^{\lambda N}(\chi) \subseteq \mathbb{M}^{N\lambda}(\chi)$. We now define a new multiplier set,

$$\widehat{\mathbb{M}}^{N\lambda}(\chi) := \left\{\widetilde{\Lambda} \in \mathbb{M}^{N\lambda}(\chi) : \widetilde{\Lambda} \text{ satisfies Eq. (1.25)}\right\} \tag{1.28}$$

Thus, $\widehat{\mathbb{M}}^{N\lambda}(\chi) \sim \mathbb{M}^{\lambda N}(\chi)$. That is, under a matching (closure) condition, every solution of Problem $B^{N\lambda}$ is also a solution to Problem $B^{\lambda N}$. This statement is encapsulated as the Covector Mapping Theorem:

Theorem 1 (Unified Covector Mapping Theorem). *Let* $\mathbb{M}^{\lambda N}(\chi) \neq \emptyset$ *and* $\{\widehat{\nu}_0, [\widehat{\mu}_k], [\widehat{\lambda}_k]\} \in \widehat{\mathbb{M}}^{N\lambda}(\chi)$*; then, the bijection,* $\widehat{\mathbb{M}}^{N\lambda}(\chi) \sim \mathbb{M}^{\lambda N}(\chi)$*, is given by,*

$$\lambda^N(t_k) = \widehat{\lambda}_k \quad \mu^N(t_k) = \widehat{\mu}_k, \quad \nu_0 = \widehat{\nu}_0 \tag{1.29}$$

Remark 2. The statement of Theorem 1 is identical to that of [63] and is made possible through the construction of weighted interpolation. The weighted interpolants lead to a consistent pair of differentiation matrices, D and D^*, that are dual to each other. In this context, the LGL case turns out to be the special situation where the formal adjoint of D is the same as $-D$.

Remark 3. Note that Theorem 1 neither states $\lambda^N(t_k) = \widetilde{\lambda}_k$, $\mu^N(t_k) = \widetilde{\mu}_k$, $\nu_0 = \widetilde{\nu}_0$ nor does it imply $\lambda^N(t_k) \approx \widetilde{\lambda}_k$, $\mu^N(t_k) \approx \widetilde{\mu}_k$, $\nu_0 \approx \widetilde{\nu}_0$ as is sometimes erroneously interpreted. Furthermore, note that Eq. (1.29) is an exact relationship.

Remark 4. A simple counter example is constructed in [45] to show that a solution to Problem $B^{\lambda N}$ may not exist for Euler discretization no matter how small the mesh. This well-known phenomenon requires a proper technical modification to Theorem 1 similar to Polak's theory of consistent approximations. This aspect of Theorem 1 is rigorously proved in [40] and [41] for unit weight functions. The extension of this rigor to non-unit weight function is straightforward but lengthy. Thus, the main assumption in Theorem 1 that the multiplier set, $\mathbb{M}^{\lambda N}(\chi)$, is non empty can be eliminated. In other words, Theorem 1 remains valid under much weaker assumptions.

Remark 5. Because the set $\mathbb{M}^{N\lambda}(\chi)$ is "larger" (i.e. has more elements) than $\mathbb{M}^{\lambda N}(\chi)$, it de-sensitizes the sensitivities associated with solving Problem B^N. This fact is exploited in [64] to design a "guess-free" algorithm for solving Problem B.

Remark 6. A satisfaction of the CMP is not a sufficient condition for convergence. In fact, it is shown in [65] that it is possible for the costates to converge even when the controls do not converge. This can happen when when chooses the LG or the LGR grid points to solve finite horizon optimal control problems.

1.5.2 Some Remarks on the Covector Mapping Principle

The Covector Mapping Theorem was generated by an application of the Covector Mapping Principle (see Fig. 1.7). Since the introduction of Pontryagin's Principle, it has been known that the Minimum Principle may fail in the discrete-time domain if it is applied in exactly the same manner as in the continuous-time domain. For the Minimum Principle to hold exactly in the discrete-time domain, additional assumptions of convexity are required whereas no such assumptions are necessary for the continuous-time versions. This is because the continuous-time problem has a *hidden convexity* [62] that is implicitly used in proving theorems. Thus, when a continuous-time optimal control problem is discretized, this hidden convexity is not carried over to the discrete time domain resulting in a loss of information. If this information loss is not restored, the discrete-time solution may be spurious, not converge to the correct solution, or may even provide completely false results. A historical account of these issues, along with a simple counter example, is described in [45]. Further details are provided in the references contained in [45]. A thorough discussion of these issues and their relationship to advance concepts in optimal control theory has been developed by Mordukhovich. [61, 62].

The closure conditions introduced by Ross and Fahroo [37] are a form of matching conditions that are similar in spirit to Mordukhovich's matching conditions for Euler approximations. These conditions can be in primal space alone [61], or in primal-dual space [46]. Note however that the primal space conditions [61] are obtained through dual space considerations. These new ideas reveal that dual space issues cannot be ignored even in the so-called "direct methods," for optimal control.

1.5.3 Software

A proper implementation of a PS method requires addressing all the numerical stability and accuracy issues and a case-by case approach to problem solving using PS approximations is inadvisable as it is tantamount to using first principles in every situation. Consequently, it is preferable for all the intricacies of a PS method to be implemented just once for a general problem in a form of a reusable software package. This exercise has been carried out in both OTIS [8] and DIDO [7]. OTIS is in FORTRAN while DIDO is in MATLAB. While OTIS uses PS techniques as one of its many option, DIDO is exclusively based on PS methods alone.

In addition, DIDO implements the spectral algorithm [66] to complete the "circuit" shown in Fig. 1.7.

1.6 Applications to Rapid UAV Trajectory Optimization

MONARC is a fixed-wing experimental UAV outfitted at the Naval Postgraduate School to test agility concepts for UAVs; see Fig. 1.8. Savage [67] has tested out

Fig. 1.8 MONARC test apparatus at the Naval Postgraduate School [67]

a number of concepts according to a work-flow plan shown in Fig. 1.9. That is, an optimal control problem is formulated as Problem B as an input to DIDO. DIDO then performs all the remaining corners illustrated in Fig. 1.9. That is, the flight implementation is done in primal space (Problem B^N) while the dual space results (Problem B^λ) are used for a pre-flight verification and validation of the solution using the equivalence result (Problem $B^{N\lambda}$ = Problem $B^{\lambda N}$) facilitated by the CMP.

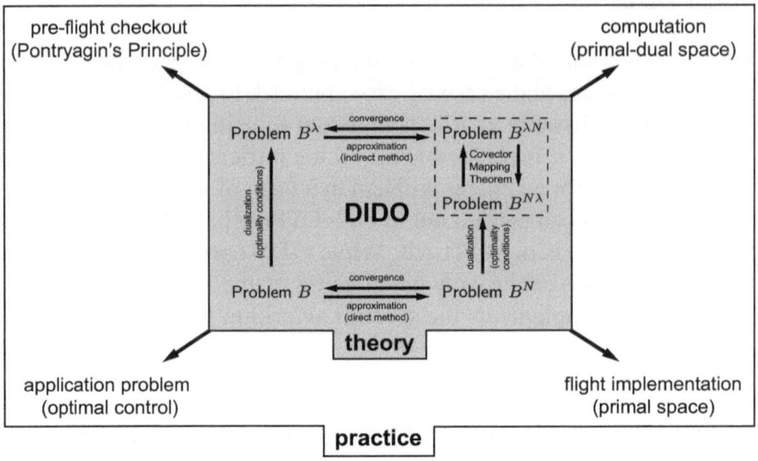

Fig. 1.9 Illustrating the work-flow and the close connection between software, theory and practice of PS optimal control; compare with Fig. 1.7

As part of the work-flow, a sequence of optimal control problems is set up with increasing model fidelity. During each stage the concepts are tested in both simulation and hardware as shown in Fig. 1.10. In this section, we describe the very first phase of this work-flow. In this phase, the kinematics of the MONARC are augmented with inertias for the velocity, v, flight path elevation, γ, and flight path heading angle, ζ, in order to generate controls for acceleration, u_a, pitch rate of change, u_γ, and heading rate of change, u_ζ. This concept generates the Phase 1 dynamics for the MONARC given by,

Fig. 1.10 Hardware-in-the-Loop apparatus at the Naval Postgraduate School [67]

$$\dot{x} = v\cos(\gamma)\cos(\zeta) \tag{1.30}$$
$$\dot{y} = v\cos(\gamma)\sin(\zeta) \tag{1.31}$$
$$\dot{z} = v\sin(\gamma) \tag{1.32}$$
$$\dot{v} = u_a \tag{1.33}$$
$$\dot{\gamma} = u_\gamma \tag{1.34}$$
$$\dot{\zeta} = u_\zeta \tag{1.35}$$

where x, y, z, denote the cartesian position coordinates of the MONARC. Thus, the state variables are, $(x, y, z, v, \gamma, \zeta) \in \mathbb{R}^6$ while the controls are $[u_a, u_\gamma, u_\zeta] \in \mathbb{R}^3$. The following bounds on the controls,

$$-0.04 \leq u_a \leq 0.04, \ -\pi/6 \leq u_\gamma \leq \pi/6, \ -\pi/6 \leq u_\zeta \leq \pi/6$$

allows us to incorporate the optimal solutions directly into the inner-loop.

A particular air combat maneuver of interest is a minimum-time velocity reversal maneuver (VRM). While a convention VRM is a post-stall Herbst maneuver [68, 69], we perform a Phase 1 analysis to keep the MONARC below stall speed; this generates a state constraint given by,

$$0.17 \leq v \leq 1.83$$

The initial and final position for the VRM are given by,

$$x(t_0) = 0 = x(t_f), \ y(t_0) = 0 = y(t_f), \ z(t_0) = 0 = z(t_f)$$

The remaining boundary conditions that define the VRM are,

$$v(t_0) = 1 = v(t_f), \ \ , \gamma(t_0) = 0 = \gamma(t_f), \ \zeta(t_0) = 0, \ \zeta(t_f) = \pi$$

These boundary conditions indicate that the only appropriate choice of the grid is LGL; hence, in the following we limit our discussions to this grid.

An XZ-view of the VRM, plotted in Fig. 1.11 shows a "tear-drop" shape that is consistent with Savage's result [67]. The optimal maneuver exploits the full airspace in the y- and z-directions as shown in Fig. 1.12. A traditional pilot or a conventional autopilot cannot design such a maneuver, particularly, when boundary conditions are changed to accommodate different combat scenarios [67]. From its 3-D view, shown in Fig. 1.13 the VRM looks like the Herbst maneuver; however, note that this is achieved without stalling the MONARC or "training" a pilot. A large collection of such maneuvers have been designed and tested by Savage [67] including the use of models and results beyond Phase 1.

As a means to further illustrate the work-flow illustrated in Fig. 1.9, we first note that it is quite straightforward to show that λ_x, λ_y and λ_z are constants. A plot of the costates from DIDO shows exactly this behavior in Fig. 1.14.

Finally, the flyability of the maneuver is checked in several steps. One of the first steps of this check is the convergence and feasibility of the solutions. Under

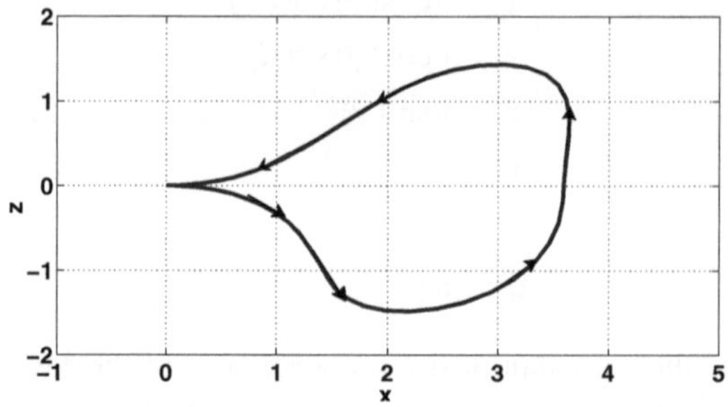

Fig. 1.11 "Tear-drop" shape of the velocity-reversal maneuver

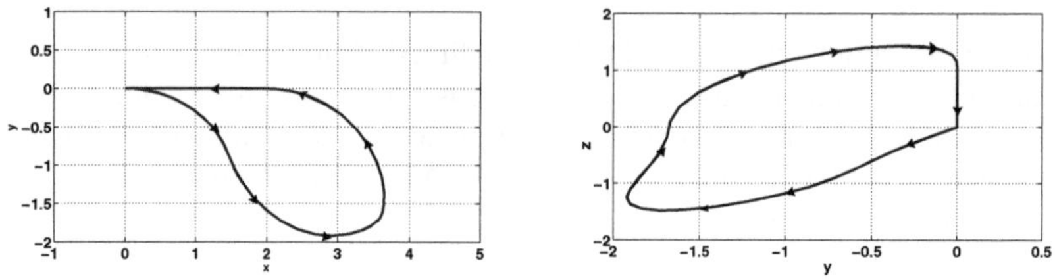

Fig. 1.12 Illustrating the exploitation of the air space via non-intuitive shaping of the VRM

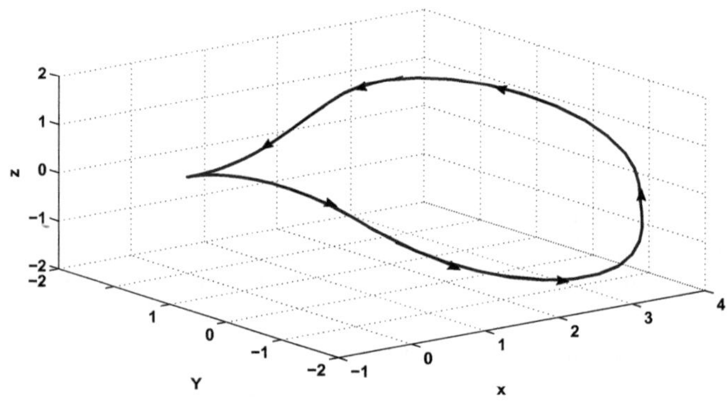

Fig. 1.13 A 3D view of VRM

this validation, the optimal controls from DIDO are used to generate an initial value problem (IVP) given by $\dot{\mathbf{x}} = \mathbf{f}(\mathbf{x}, \mathbf{u}(t), t)$, $\mathbf{x}(t_0) = \mathbf{x}^0$, and the IVP is solved using a standard propagator such as ODE45 in MATLAB. The results of this validation are shown in Fig. 1.15. In this figure, the states from the LGL grid are superimposed on the same graph of the states obtained from ODE45. Clearly, the results are visually indistinguishable from each other. These primal solutions are part of the flight

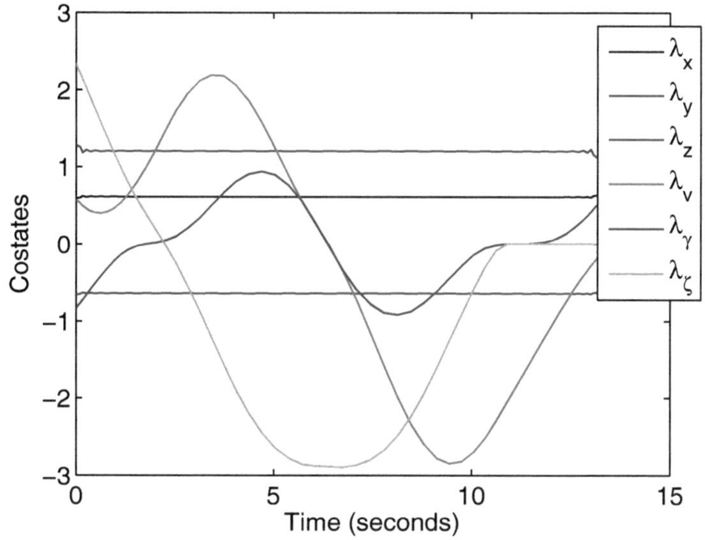

Fig. 1.14 Costates for the VRM obtained from DIDO and illustrating the work-flow of the top left corner of Fig. 1.9

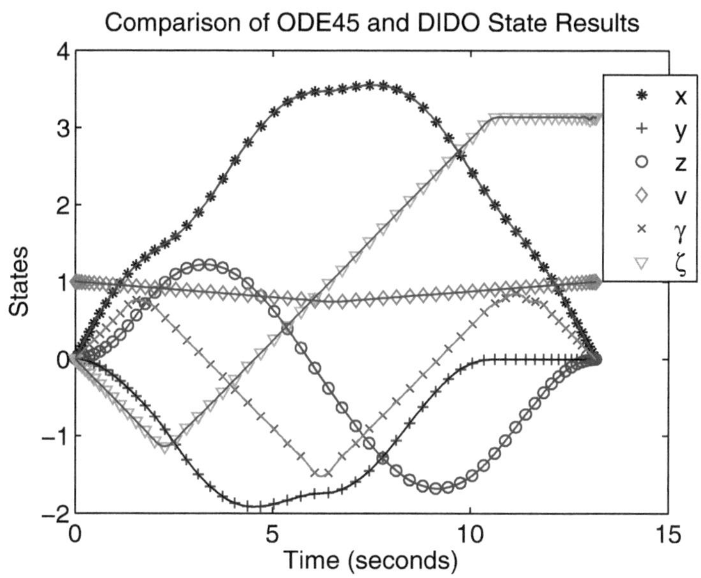

Fig. 1.15 Illustrating first tests on the flyability of PS controls obtained from DIDO: solid lines are ODE45 propagation of PS controls and symbols are DIDO states results

implementable solutions indicated in the bottom right of Fig. 1.9. A hardware-in-the-loop validation of such solutions has been performed by Savage using the apparatus shown in Fig. 1.10.

1.7 Conclusions

In order to support a wide exposition of PS optimal control and its applications to rapid UAV motion planning, we have taken the liberty of describing our core ideas in terms of certain simplifications. As we briefly noted in Sec. 1.2, this is not to be confused as a limitation of the ideas; rather, with additional bookkeeping all our ideas transfer trivially to substantially more complex problems than the one posed as Problem B. In the same spirit, we have restricted the discussions in this paper to Legendre-based PS grids. The key concept we have exposed in this paper is contained in Eq. (1.18). This is the notion of consistent primal and dual weight functions for interpolation. Roughly speaking, PS concepts for optimal control can be described as follows: For any chosen primal interpolant with primal weight function, $W(t)$, it is necessary to choose a consistent dual interpolant with weight function, $W^*(t)$. The weight function pair, $\{W(t), W^*(t)\}$, must be selected in a manner that generates the correct Hilbert space for the approximation of functions. Under the assumption of existence of a continuous-time solution in an appropriate Sobolev space, global convergence of the solution based on the consistency of the approximation can then be rigorously proved under mild and checkable conditions. References [40, 41, 70, 71, 72] present some of the foundations for such results. These theoretical developments explain the consistent performance of the correct PS methods for optimal control. That is, is is possible to inadvertently design a disastrous PS technique (such as the one based on a pure Gauss grid or an incorrect application of a Radau family) leading to a flight failure. A wide exposition of these ideas are summarized in [1].

References

1. Ross, I.M., Karpenko, M.: A review of pseudospectral optimal control: from theory to flight. Annual Reviews in Control 36, 182–197 (2012)
2. Kang, W., Bedrossian, N.: Pseudospectral optimal control theory makes debut flight – saves NASA $1M in under 3 hrs. SIAM News 40(7), 1 (2007)
3. Gong, Q., Kang, W., Bedrossian, N., Fahroo, F., Sekhavat, P., Bollino, K.: Pseudospectral Optimal Control for Military and Industrial Applications. Special Session of the 46th IEEE Conference on Decision and Control, New Orleans, LA, pp. 4128–4142 (December 2007)
4. Bedrossian, N., Bhatt, S., Kang, W., Ross, I.M.: Zero-propellant Maneuver Guidance. IEEE Control Systems Magazine, 53–73 (October 2009)
5. Keesey, L.: TRACE Spacecraft's New Slewing Procedure (December 20, 2010), http://www.nasa.gov/mission_pages/sunearth/news/trace-slew.html

6. Bedrossian, N., Karpenko, M., Bhatt, S.: Overclock My Satellite: Sophisticated algorithms boost satellite performance on the cheap. IEEE Spectrum (November 2012)
7. Ross, I.M.: A Beginner's Guide to DIDO: A MATLAB Application Package for Solving Optimal Control Problems. Elissar Global, Monterey (2007)
8. Riehl, J., Paris, S., Sjauw, W.: OTIS 4.0 User Manual, NASA Glenn Research Center, Cleveland, OH (2008)
9. Lu, P., Sun, H., Tsai, B.: Closed-loop endoatmospheric ascent guidance. Journal of Guidance, Control and Dynamics 26(2), 283–294 (2003)
10. Melton, R.G.: Comparison of direct optimization methods applied to solar sails. In: AIAA/AAS Astrodynamics Specialists Conference and Exhibit, Monterey, CA, August 5-8, AIAA 2002-4728 (2002)
11. Paris, S.W., Riehl, J.P., Sjaw, W.K.: Enhanced procedures for direct trajectory optimization using nonlinear programming and implicit integration. In: AIAA/AAS Astrodynamics Specialist Conference and Exhibit, Keystone, CO, August 21-24, AIAA 2006-6309 (2006)
12. Riehl, J.P., Paris, S.W., Sjaw, W.K.: Comparison of implicit integration methods for solving aerospace trajectory optimization problems. In: AIAA/AAS Astrodynamics Specialist Conference and Exhibit, Keystone, CO, August 21-24, AIAA 2006-6033 (2006)
13. Bedrossian, N., Bhatt, S., Lammers, M., Nguyen, L., Zhang, Y.: First ever flight demonstration of zero propellant maneuver attitude control concept. AIAA-2007-6734
14. Bedrossian, N., Bhatt, S., Lammers, M., Nguyen, L.: Zero propellant maneuver: flight results for 180° ISS rotation. In: 20th International Symposium on Space Flight Dynamics, September 24-28, Annapolis, MD, NASA/CP-2007-214158 (2007)
15. Harada, M., Bollino, K.: Optimal trajectory of a glider in ground effect and wind shear. In: AIAA Guidance, Navigation and Control Conference, San Francisco, CA, August 15-18, AIAA 2005-6474 (2005)
16. Harada, M., Bollino, K., Ross, I.M.: Minimum-fuel circling for an unmanned aerial vehicle. In: 2005 JSASS-KSAS Joint International Symposium on Aerospace Engineering, Nagoya, Japan, October 12-15 (2005)
17. Hawkins, A.M., Fill, T.R., Proulx, R.J., Feron, E.M.: Constrained trajectory optimization for lunar landing. In: AAS Spaceflight Mechanics Meeting, Tampa, FL, AAS 06-153 (January 2006)
18. Infeld, S.I., Murray, W.: Optimization of stationkeeping for a libration point mission. In: AAS Spaceflight Mechanics Meeting, Maui, HI, AAS 04-150 (February 2004)
19. Pietz, J., Bedrossian, N.: Momentum dumping using only CMGs. In: Proceedings of the AIAA GNC Conference, Austin, TX (2003)
20. Rea, J.: Launch vehicle trajectory optimization using a Legendre pseudospectral method. In: Proceedings of the AIAA Guidance, Navigation and Control Conference, Austin, TX, AIAA 2003-5640 (August 2003)
21. Stanton, S., Proulx, R., D'Souza, C.: Optimal orbit transfer using a Legendre pseudospectral method. In: AAS/AIAA Astrodynamics Specialist Conference, Big Sky, MT, August 3-7, AAS-03-574 (2003)
22. Williams, P.: Application of pseudospectral methods for receding horizon control. J. of Guid., Contr. and Dyn. 27(2), 310–314 (2004)
23. Williams, P., Blanksby, C., Trivailo, P.: Receding horizon control of tether system using quasilinearization and Chebyshev pseudospectral approximations. In: AAS/AIAA Astrodynamics Specialist Conference, Big Sky, MT, August 3-7, AAS 03-535 (2003)
24. Stevens, R.E., Wiesel, W.: Large time scale optimal control of an electrodynamic tether satellite. Journal of Guidance, Control and Dynamics 32(6), 1716–1727 (2008)

25. Williams, P.: Three-dimensional aircraft terrain-following via real-time optimal control. Journal of Guidance, Control and Dynamics 30(4), 1201–1205 (2007)
26. Yan, H., Alfriend, K.T.: Three-axis magnetic attitude control using pseudospectral control law in eccentric orbits. In: AAS Spaceflight Mechanics Meeting, Tampa, FL, AAS 06-103 (January 2006)
27. Li, J.-S., Ruths, J., Yu, T.-Y., Arthanari, H., Wagner, G.: Optimal pulse design in quantum control: a unified computational method. Proceedings of the National Academy of Sciences 108(5), 1879–1884 (2011)
28. Ross, I.M., Fahroo, F.: Pseudospectral methods for optimal motion planning of differentially flat systems. IEEE Trans. on Automat. Contr. 49(8), 1410–1413 (2004)
29. Bollino, K., Lewis, L.R., Sekhavat, P., Ross, I.M.: Pseudospectral optimal control: a clear road for autonomous intelligent path planning. In: Proc. of the AIAA Infotech Aerospace 2007 Conference, CA (May 2007)
30. Bollino, K., Lewis, L.R.: Optimal path planning and control of tactical unmanned aerial vehicles in urban environments. In: Proc. of AUVSI's Unmanned Systems North America 2007 Conference, Washington D.C. (August 2007)
31. Bollino, K.P., Lewis, R.L.: Collision-free multi-UAV optimal path planning and cooperative control for tactical applications. In: Proceedings of the AIAA Guidance, Navigation and Control Conference and Exhibit, Honolulu, Hawaii, August 18-21 (2008)
32. Lewis, L.R.: Rapid motion planning and autonomous obstacle avoidance for unmanned vehicles. Masters Thesis, Naval Postgraduate School, Monterey, CA (December 2006)
33. Fahroo, F., Ross, I.M.: On discrete-time optimality conditions for pseudospectral methods. In: Proceedings of the AIAA/AAS Astrodynamics Conference, Keystone, CO, AIAA-2006-6304 (August 2006)
34. Elnagar, J., Kazemi, M.A., Razzaghi, M.: The pseudospectral Legendre method for discretizing optimal control problems. IEEE Transactions on Automatic Control 40(10), 1793–1796 (1995)
35. Fahroo, F., Ross, I.M.: Costate estimation by a Legendre pseudospectral method. Journal of Guidance, Control and Dynamics 24(2), 270–277 (2001)
36. Ross, I.M., Fahroo, F.: A pseudospectral transformation of the covectors of optimal control systems. In: Proceedings of the First IFAC Symposium on System Structure and Control, Prague, Czech Republic, August 29-31 (2001)
37. Ross, I.M., Fahroo, F.: Legendre pseudospectral approximations of optimal control problems. In: Kang, W., Borges, C., Xiao, M. (eds.) New Trends in Nonlinear Dynamics and Control. LNCIS, vol. 295, pp. 327–342. Springer, Heidelberg (2003)
38. Fahroo, F., Ross, I.M.: Direct trajectory optimization by a Chebyshev pseudospectral method. Journal of Guidance, Control and Dynamics 25(1), 160–166 (2002)
39. Gong, Q., Ross, I.M., Fahroo, F.: Chebyshev pseudospectral method for nonlinear constrained optimal control problems. In: 48th IEEE Conference on Decision and Control, Shanghai, China, pp. 5057–5062 (December 2009)
40. Gong, Q., Kang, W., Ross, I.M.: A pseudospectral method for the optimal control of constrained feedback linearizable systems. IEEE Transactions on Automatic Control 51(7), 1115–1129 (2006)
41. Gong, Q., Ross, I.M., Kang, W., Fahroo, F.: On the pseudospectral covector mapping theorem for nonlinear optimal control. In: Proceedings of the 45th IEEE Conference on Decision and Control, San Diego, CA, December 13-15 (2006)
42. Gong, Q., Ross, I.M., Kang, W., Fahroo, F.: Connections between the covector mapping theorem and convergence of pseudospectral methods for optimal control. Computational Optimization and Applications 41, 307–335 (2008)

43. Ross, I.M.: A Primer on Pontryagin's Principle in Optimal Control. Collegiate Publishers, San Francisco (2009)
44. Ross, I.M., Fahroo, F.: A perspective on methods for trajectory optimization. In: Proceedings of the AIAA/AAS Astrodynamics Conference, Monterey, CA, AIAA Paper No. 2002-4727 (August 2002)
45. Ross, I.M.: A historical introduction to the covector mapping principle. In: Advances in the Astronautical Sciences, Univelt, San Diego, CA, vol. 123, pp. 1257–1278 (2006)
46. Ross, I.M.: A road map for optimal control: the right way to commute. Annals of the New York Academy of Sciences 1065 (January 2006)
47. Ross, I.M.: Certain connections in optimal control theory and computation. In: Proceedings of the Symposium on New Trends in Nonlinear Dynamics and Control and Their Applications, Monterey, CA, October 18-19 (2002) (invited talk)
48. Davis, P., Rabinowitz, P.: Methods of Numerical Integraion. Academic Press (1975)
49. Karniadakis, G., Sherwin, S.: Spectr al/hp Element Methods for Computational Fluid Dynamics. Oxford University Press, Oxford (2005)
50. Fahroo, F., Ross, I.M.: Pseudospectral methods for infinite horizon nonlinear optimal control problems. In: AIAA Guidance, Navigation and Control Conference, San Francisco, CA (2005)
51. Ross, I.M., Gong, Q., Fahroo, F., Kang, W.: Practical stabilization through real-time optimal control. In: 2006 American Control Conference, Minneapolis, MN, June 14-16 (2006)
52. Ross, I.M., Fahroo, F.: Discrete verification of necessary conditions for switched nonlinear optimal control system. In: Proceedings of the American Control Conference, Boston, MA (June 2004)
53. Fahroo, F., Ross, I.M.: Pseudospectral methods for infinite horizon optimal control problems. Journal of Guidance, Control and Dynamics (July-August 2008)
54. Fahroo, F., Ross, I.M.: Advances in pseudospectral methods for optimal control. In: AIAA Guidance, Navigation, and Control Conference, AIAA Paper 2008-7309 (2008)
55. Gong, Q., Ross, I.M., Fahroo, F.: Pseudospectral optimal control on arbitrary grids. In: AAS/AIAA Astrodynamics Specialist Conference, Pittsburgh, PA, AAS 09-405 (2009)
56. Canuto, C., Hussaini, M.Y., Quarteroni, A., Zang, T.A.: Spectral Methods in Fluid Dynamics. Springer, New York (1988)
57. Trefethen, L.N.: Spectral Methods in MATLAB. SIAM, Philadelphia (2000)
58. Betts, J.T.: Practical Methods for Optimal Control Using Nonlinear Programming. SIAM, Philadelphia (2001)
59. Ross, I.M., Gong, Q., Sekhavat, P.: Low-thrust, high-accuracy trajectory optimization. Journal of Guidance Control and Dynamics 30(4), 921–933 (2007)
60. Ross, I.M., Gong, Q., Sekhavat, P.: The Bellman pseudospectral method. In: AIAA/AAS Astrodynamics Specialist Conference and Exhibit, Honolulu, Hawaii, August 18-21, AIAA-2008-6448 (2008)
61. Mordukhovich, B.S.: Variational Analysis and Generalized Differentiation, I: Basic Theory. Grundlehren der Mathematischen Wissenschaften [Fundamental Principles of Mathematical Sciences], vol. 330. Springer, Berlin (2005)
62. Mordukhovich, B.S.: Variational Analysis and Generalized Differentiation, II: Applications. Grundlehren der Mathematischen Wissenschaften [Fundamental Principles of Mathematical Sciences], vol. 331. Springer, Berlin (2005)
63. Ross, I.M., Fahroo, F.: Discrete verification of necessary conditions for switched nonlinear optimal control systems. In: Proceedings of the American Control Conference, Boston, MA (June 2004)

64. Ross, I.M., Gong, Q.: Guess-free trajectory optimization. In: AIAA/AAS Astrodynamics Specialist Conference and Exhibit, Honolulu, Hawaii, August 18-21, AIAA Paper 2008-6273 (2008)
65. Fahroo, F., Ross, I.M.: Convergence of the costates does not imply convergence of the control. Journal of Guidance, Control, and Dynamics 31(5), 1492–1497 (2008)
66. Gong, Q., Fahroo, F., Ross, I.M.: A spectral algorithm for pseudospectral methods in optimal control. AIAA Journal of Guidance, Control and Dynamics 31(3), 460–471 (2008)
67. Savage, M.K.: Design and hardware-in-the-loop implementation of optimal canonical maneuvers for an autonomous planetary aerial vehicle. M.S. thesis, Dept. of Mech. and Aerospace Eng., Naval Postgraduate School, Monterey, CA (2012)
68. Carter, B.R.: Time-optimization of high performance combat maneuvers. M.S. thesis, Dept. of Mech. and Astronautical Eng., Naval Postgraduate School, Monterey, CA (2006)
69. Fabey, M.J., Fulghum, D.A.: Turn and burn: Raptor's gunfighting heritage hasn't completely disappeared. Aviation Week & Space Technology, 4–5 (January 8, 2007)
70. Kang, W.: Rate of convergence for the Legendre pseudospectral optimal control of feedback linearizable systems. Journal of Control Theory and Application 8(4), 391–405 (2010)
71. Kang, W., Ross, I.M., Gong, Q.: Pseudospectral optimal control and its convergence theorems. In: Analysis and Design of Nonlinear Control Systems, pp. 109–126. Springer, Heidelberg (2008)
72. Kang, W., Gong, Q., Ross, I.M.: Convergence of pseudospectral methods for a class of discontinuous-control nonlinear optimal problems. In: Proceedings of the 44th IEEE Conference on Decision and Control, Seville, Spain, December 12-15 (2005)

Chapter 2
Max Plus Decision Processes in Planning Problems for Unmanned Air Vehicle Teams

Ben G. Fitzpatrick

Abstract. Many aspects of unmanned air vehicle (UAV) operations In this paper, we consider some idempotent modifications of stochastic and Bayesian analysis suitable for decision making in mixed initiative and multi-agent systems. Of particular interest are techniques for mixed initiative and highly autonomous operation. We examine certain Markov Decision Processes (MDPs) within the context of max-plus probability, and we discuss their application to problems of control of unmanned air sensing assets.

2.1 Introduction

As the use of unmanned air vehicles (UAVs) and unmanned air systems (UAS) grows, problems of mission planning and execution become more complex. Current operations generally involve multiple human operators controlling a single vehicle, in a battlespace in which no other team assets are operating. This state of affairs is quite a distance from future Air Force plans for UAVs. In [11], the Air Force expresses a goal of moving from "man in the loop" to "man on the loop," an aspiration requiring a greatly enhanced capability in both algorithms and computations. Significant increases in autonomous capability are necessary to improve efficiency and reduce man power-related costs [12]. Autonomy is sought not only in wider use of autonomous systems but also in the degree of autonomy. In particular, the desire to move from control to autonomy ([12]) leads to some major challenges:

> Airborne remote-piloted systems and many non-airborne systems have significant autonomous capability, however increasing levels of "flexible" autonomy will be needed for a wider range of Air Force functions during 2010-2030 and beyond. These will include fully unattended systems with reliable levels of autonomous functionality far greater than is possible today, as well as systems that can reliably make wide-ranging

Ben G. Fitzpatrick
Loyola Marymount University and Tempest Technologies, Los Angeles, CA
e-mail: `bfitzpatrick@lmu.edu`

F. Fahroo et al. (Eds.): *Recent Adv. in Res. on Unmanned Aerial Veh.*, LNCIS 444, pp. 31–45.
DOI: 10.1007/978-3-642-37694-8_2

> autonomous decisions at cyber speeds to allow reactions in time-critical roles far exceeding what humans can possibly achieve.
>
> Key attributes of such autonomy include the ability for complex decision making, including autonomous mission planning, and the ability to self-adapt as the environment in which the system is operating changes. Flexible autonomy refers to systems in which a user can specify the degree of autonomous control that the system is allowed to take on, and in which this degree of autonomy can be varied from essentially none to near or complete autonomy.
>
> Limited autonomy is already being widely used in Air Force systems, but their autonomous functionality is far below what could be achieved over the next decade. From a purely technical perspective, methods already exist to extend autonomy to far higher levels than are being used today. Even this level of autonomy will be enabled only when automated methods for verification and validation (V&V) of complex, adaptive, highly nonlinear systems are developed. In the near- to mid-term, developing methods for establishing "certifiable trust in autonomous systems" is the single greatest technical barrier that must be overcome to obtain the capability advantages that are achievable by increasing use of autonomous systems.

The UASF plan for UAS [11] suggests a strong desire for a high level of autonomy:

> Much like a chess master can outperform proficient chess players, UAS will be able to react at these speeds and therefore this loop moves toward becoming a "perceive and act" vector. Increasingly humans will no longer be "in the loop" but rather "on the loop" – monitoring the execution of certain decisions. Simultaneously, advances in AI will enable systems to make combat decisions and act within legal and policy constraints without necessarily requiring human input.

While there are a number of technological difficulties in achieving this vision, automated decision processes are clearly among the most difficult. The use of computational decision aids for military purposes has had mixed results [21]. One key step in operating with a high level of trustworthiness is the development of decision rules that can be understood with sufficient clarity by operators and commanders. Humans tend to make decisions in a manner that is very different from the computational control algorithms used to determine optimal courses of action. While traditional mathematical approaches of dynamical systems models and Markov chain models are very well suited to physical problems such as flight control, they are difficult to implement and validate for higher level decision making.

One interesting and productive approach to autonomous decision making is based on Bayesian analysis. Bayesian analyses bases offer a logical and computational structure that is amenable to V& V and adaptable to uncertainties [10, 31, 23, 35]. Bayesian networks, for example, encode conditional probability structure in a causal manner to model autonomous reasoning and decision processes. Among the relevant applications of Bayesian networks to planning for autonomous UAV systems are path planning ([32, 33]), surveillance ([7, 24]), and information fusion in command and control ([5]).

In the present paper, we develop a max-plus approach to Bayesian analysis and MDPs. The max-plus probability assignment is interpreted as the cost of an event or decision rather than the likelihood of occurrence ([2, 34, 17]), making the approach well-suited to optimization and dynamic programming problems. We begin the paper with a model problem, followed by a brief review of naturalistic decision processes that humans use in command and control, Bayesian analyses, followed by the max-plus construction and a UAV example problem.

2.2 Perimeter Surveillance Example

To illustrate problems of autonomous UAV control, we consider a decision process for a UAV performing perimeter surveillance [8]. A protected site is surrounded with simple sensors that detect possible intrusion. Figure 2.1 illustrates this problem. This problem is based on the the Cooperative Operations in UrbaN TERrain (COUNTER) scenario [8, 20], in which a single operator controls four UAVs performing surveillance. The operator looks for targets in the video streams provided by the UAV. The operator tasks the UAVS to loiter for better looks or move on to the next potential target. In order to reduce operator work load, we seek a semi-autonomous support solution. The simplest first problem, as posed in [8], is to position the simple detection sensors to request UAV support. In the fully autonomous problem posed in [8, 20], the UAVs use an operator error model to decide how many loiters are needed to get a high quality detection decision. In this problem, the UAV flies around the perimeter, servicing the sensors that broadcast detections, in an attempt to verify the intruder. The simple sensors have a significant false alarm rate. The UAVs actions are to loiter at a site or to move on (unidirectionally)

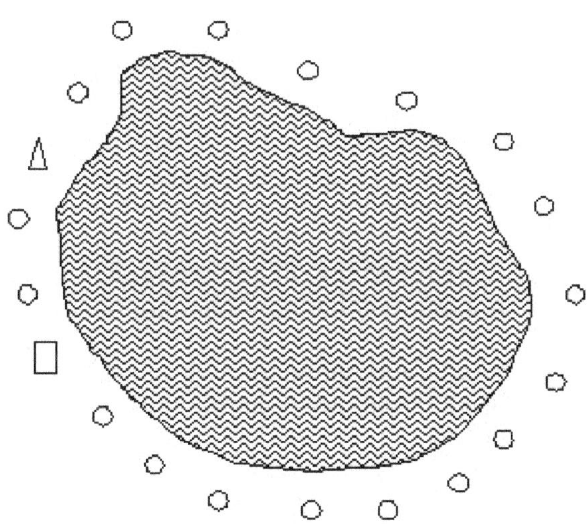

Fig. 2.1 A (shaded) protected area with intrustion detection sensors (circles), a patrolling UAV (triangle), and an intruder (rectangle)

around the perimeter. Formulations of this problem have focused on a penalized optimization criterion that maximizes information gained by loitering while penalizing unchecked alerts. More complex, realistic formulations must take into account the "man on the loop" issue.

The model proposed by Chandler and colleagues [8, 20] involves a controlled discrete dynamical system with time step k and with states $x(k) \in \{1,2,\cdots,N\}$ denoting the position along the perimeter discretized by the sensor station, $\tau_i, i \in \{1,2,\cdots,N\}$ denoting the length of time station i has been on alert, $S_i(k), i \in \{1,2,\cdots,N\}$ denoting the (0,1) alert status of station i, and $d(k)$ denoting the dwell time of the UAV at its current location. We have exogenous variables $Y_i(k)$ denoting the alert arrival indicator of station i at time k. These arrivals may be threats, with probability p, or nuisances with probability $1-p$. We also have $Z(k)$ denoting a threat detection at time k at the UAV's current location. The threat detection is a random quantity modeled as an operator decision under uncertainty, typically involving the operator viewing video stream from the UAV. The longer the dwell time, the more likely the decision is to be correct. We model the process as follows:

$$\begin{aligned}
x(k+1) &= (x(k)+u(k))\mathrm{mod}(N),\\
\tau_i(k+1) &= (\tau_i(k)+\Delta t)S_i(k), i=1,2,\cdots,N,\\
S_i(k+1) &= (1-\delta_{i,x(k)})\max(S_i(k),Y_i(k)), i=1,2,\cdots,N,\\
d(k+1) &= (d(k)+1)\delta_{u(k),0},
\end{aligned}$$

in which $\delta_{i,j}$ denotes the Kronecker delta, and the mod operator models the fact that the perimeter is a closed curve. The optimization criterion suggested (and implemented) in [8] is given by

$$J(u) = E\Big[\sum_{k=1}^{\infty} \lambda^k (I(x(k),d(k),u(k)) - \beta \max_i(\tau_i(k)),$$

in which ΔI denotes the information gained from observing the alert stations. In order to define the information gain, we must consider the operator model for detection. We define

$$\begin{aligned}
P_{CT}(d) &= a_T + b_T(1-\exp(-\mu_T d)\\
P_{CN}(d) &= a_N + b_N(1-\exp(-\mu_N d)
\end{aligned}$$

as the probabilities of correctly detecting a target CT and nuisance CN with a dwell time of d. With these, Chandler and colleagues [8, 20] define an (Shannon) information gain in d dwells as

$$\begin{aligned}
I(d) = {} & pP_{CT}\log\Big(\frac{P_{CT}}{pP_{CT}+(1-p)(1-P_{CN})}\Big)\\
&+p(1-P_{CT})\log\Big(\frac{1-P_{CT}}{p(1-P_{CT})+(1-p)(1-P_{CN})}\Big)
\end{aligned}$$

$$(1-p)(1-P_{CN})\log\Big(\frac{1-P_{CN}}{pP_{CT}+(1-p)(1-P_{CN})}\Big)$$
$$+(1-p)P_{CN}\log\Big(\frac{P_{CN}}{p(1-P_{CT})+(1-p)(1-P_{CN})}\Big).$$

This information gain is an increasing, concave function (as illustrated in [20]), and hence there would appear to be a point of diminishing returns for dwelling versus moving on to another alert. In order to build a Bayesian network, we develop a slightly different decision criterion, which we will discuss below. An important issue in dealing with mixed initiative problems with humans and automated decision systems working together is the manner in which humans make decisions. In the following section, we review some relevant literature on human decision processes and their relationships with automated ones.

2.3 Human Decision Making and Computational Decisions

Many if not most quantitative approaches to command and control involve the specification of an optimality criterion, called an objective function, utility function, or cost function. If the problem is a dynamic one, requiring control variables to be specified over time, apply standard techniques of dynamic programming or optimal control. When humans are directly in the loop, however, they do not typically think in terms of optimizing an objective function.

For certain very low-level autonomous tasks, computational control methods work very well. Autopilots, for example, have become quite advanced and are highly effective, allowing a remote pilot to provide minimal steering control input. Generally speaking, the level of interaction is supervision: the pilot may switch off the autopilot and take full control or merely observe that the UAV is on a correct course. However, once the level of autonomy increases to the point at which a true collaboration between human and machine is required, such as target selection and the decision to fire (that is, "perceive and act" [11]), the problem becomes much more challenging.

Kessel [21] notes that problematic interactions between humans and computational decision algorithms generally arise due to a lack of trust. For the most part, the lack of trust arises from opacity of the computational decision process to the human "on the loop."

> Taken at face value, the unadorned automated choice selection no more constitutes advice than a mystical oracles bare directive. As Polanyi (1962; pp.28) points out, "An unasserted sentence is no better than an unsigned check; just paper and ink without power or meaning." The advice has a detached, unreal quality. It calls for trust or distrust, not for expert deliberation. The automatic decision may have been expertly derived from hidden information, or, what is more likely in complex decisions, it may have been drawn from mistaken perceptions, overlooked information, or simplistic heuristics. Decision makers simply do not know. They must instead make a willful resolution either to trust or to distrust the computer advice, just as those who consult

> a mystical oracle must do when they do not have contextual information about the character, reliability, and interests of the advisor.

Linegang and colleages [26] and Woods [40] also note the challenges of autonomy in mixed initiative systems, particularly humans controlling teams of UAVs. Again, a major problem is poor feedback of the autonomous system's action choice. Woods [40] suggests that humans interacting with autonomous machine agents must develop a mental model to anticipate the machine's behavior (much like humans do with other humans in team activities):

> Effective feedback depends on more than display formats; it is a relation between the systems function, the image the system presents to outsiders, and the observer embedded in an evolving context. As a result, it is better to refer to interface and feedback issues in terms of observability. This term captures the fundamental relationship between thing observed, observer and context of observation that is fundamental to effective feedback. Observability depends on the cognitive work needed to extract meaning from the data available. We, as researchers, need to make progress on better ways to measure this property of cognitive systems.

The notion of observability here is more like the systems concepts of identifiability and estimability: the human on the loop is attempting to understand at some level the feedback controller of the machine.

To reconcile algorithmic approaches to decision and control with the needs of mixed initiative problems and autonomy, research into human decision processes provides some important information. Recent research into human decision making (see, e.g., [22, 25]) suggests that humans tend to choose actions to match situations rather than to compare alternative actions to find the "best" one. This type of decision process is referred to as naturalistic decision making (NDM). Simon's bounded rationality work [36] indicates human decision makers are "satisficers" rather than "optimizers," seeking a solution that works rather than the best among all possible choices. Recognition Primed Decision Making (RPD) is a form of bounded rationality thinking, a pattern matching or feedback estimation that compares the situation to known or previously experienced situations. The decision rule then is to apply the decision rule that "worked" in the previous experience. RPD is a key characteristic of proficient decision makers. One can view this process as a single pass through an OODA (Observe, Orient, Decide, Act) loop [30].

Of course, an open loop, one time decision is rarely meaningful in complex situations such as the management of teams of unmanned and manned systems in the battlefield. Klein [22] notes that proficient decision makers perform mental simulation in the process of RPM to form the initial recognition, to generate expectations to verify the recognition, and to evaluate the selected course of action. The closing of the loop, and the iteration of the process, brings us to the consideration of uncertainty.

Lipshitz and Strauss [27] identified three sources of uncertainty: lack of information, inadequate understanding, and conflicted alternatives. Lack of information is the traditionally modeled "partially observed" uncertainty in many quantitative decision and control problems. Inadequate information can range from a lack of state identifiability to an inadequte pattern-matching set (or incomplete model). Conflicting alternatives also falls into the inadequately modeled analogy in that the decision maker cannot differentiate among the effects to be acheived by at least two possible actions. The response to uncertainty generally depends on the type of uncertainty: lack of information is handled by collecting more, assumption-based reasoning is used to cope with inadequate understanding, and weighing of pros and cons is applied to treat conflicting alternatives.

While researchers proposing the NDM model of human command and control suggest that informal, qualitative, and empirical methods are most suited to its analysis, we feel that it parallels quite closely certain Bayesian decision theoretic approaches. That is, one approaches a situation with some prior information, the patterns to which the situation will be matched. Having performed what is essentially, if not quantitatively, a likelihood estimation, one selects an action that performed for the estimated state. The primary distinction between the NDM model and more traditional optimization based approaches is in the specification of the objective. Generally speaking, human decision makers appear to evaluate possible actions in terms of "will it work?" rather than its proximity to a continuous definition of optimality [22]. This leads us to interpret such human-based command and control problems more in terms of a success-failure objective rather than a traditional utility or cost function. This sort of decision criterion actually lends itself well to a Bayesian view, in which the decision maker seeks from experience (his or her prior) the action that will have the best probability of success.

We should note here that, in addition to the quite applied research behind NDM, there has been a significant amount of more theoretical research into human decisions. The emergence of Prospect Theory and Cumulative Prospect Theory (see, e.g., [14, 18, 19, 39] as a means of dealing with bounded rationality (the limited ability of humans to collect and process information for decision making). Kahneman and Tvserky were awarded the Nobel Prize in Economics for their work in this area, in particular the demonstration that people prefer certainty to uncertain outcomes with higher expected utility and its explanation with an editing phase and an evaluation phase, processes closely aligned with NDM models.

On the other hand, it is also notable that experimentalists have found compelling evidence (see, e.g., [3, 13, 16, 28, 38]) that people frequently operate as if performing a Bayesian computation. These results often involve prediction and estimation, as opposed to decision making. It should also be noted that most of these studies do not offer a situation in which the human test subjects have to cope with very unlikely events or events occuring in the relatively distant future. In such situations, the observations of Kahneman and Tverksy indicate that humans poorly estimate both the very likely and very unlikely. Humans tend to overweight small probabilities and underweight larger ones.

Our goal here is in developing control and optimization algorithms that present humans on the loop with appropriate tools to integrate computational decisions from an applied point of view. In order to do so, the algorithms must respect human decision processes to a level that humans develop trust in the results. Algorithms that provide feedback that is "human modelable" are of primary interest. Toward that end, Kessel [21] notes that Bayesian and belief networks offer a means for the computational algorithm to be queried by a human regarding its decision path.

In the present paper, we examine a particular class of Bayesian tools for decision making, based on Bayesian networks and Bayesian knowledge bases (introduced in the next section). We follow these with a max-plus interpretation that in a sense amplifies small probabilities and dampens larger ones and thus has the potential to correspond more closely to human decision making.

2.4 Bayesian Approaches to Control and Optimization

Bayesian decision theory [4, 10, 31] involves a few basic ingredients. First is a prior distribution q on a state space of interest. The second is a measurement model $f(y|x)$ that denotes the likelihood of the measurement y given the true state is x. Bayes' rule provides the conditional distribution of x given y: it takes the form

$$q(x|y) = \frac{f(y|x)q(x)}{\sum_{x'} f(y|x')q(x')}$$

in discrete spaces.

Decision theory requires a decision or control variable, u, which enters through the likelihood: $f(y|x,u)$. The mathematical structure of Bayesian decision theory is used in the modeling of rational human decision making, in which a utility function $C = C(u|x)$ is used to gauge the values of decisions given the true state. The expected utility $L(u|y)$ is given by

$$L(u|y) = \sum_x C(u|x)q(x|y,u).$$

Traditional economic theory is that people make choices by maximizing such an expected utility, while prospect theory and its relatives weight the probabilities according to observations on human behavior. NDM generally makes no formal connection to Bayesian decision theory, but there are ways to view NDM as a Bayesian process. In particular, when the utility function C denotes a success/failure or pass/fail requirement, the types of decision problems of interest in NDM can be viewed as maximizing utility. The algorithm which the decision maker applies, however, is not a typical search algorithm but more of a pattern match in which previous solutions are applied and adjusted.

Many problems of interest have structure that simplifies a Bayesian computation in the presence of higher dimensions of state, measurement, and control variables. Bayesian networks provide additional analytic tools in certain cases. Moreover, the network structure of the approach lends itself well to the pattern matching "computation" humans make in the NDM model. As noted in Section 2.3, we are primarily interested in algorithmic approaches to which human decision making can be related. Toward that end, we consider Bayesian networks.

A Bayesian network involves several ingredients: a graph (or network), a random variable for each vertex (or node), and a conditional probability function that respects the network in a particular way. We begin with a directed acyclic graph (V,E) of vertices V and edges E. We denote by $\mathbf{pa}(v)$ the "parent" vertices of the node v:

$$\mathbf{pa}(v) = \{w \in V\colon (w,v) \in E\}.$$

We denote by $\mathbf{pa}(V)$ the set of parent sets:

$$\mathbf{pa}(V) = \{\mathbf{pa}(v)\colon v \in V\}.$$

We also require a standard probability space $(\Omega, \mathscr{F}, P)$ on which we have random variables $X_v, v \in V$. The final component of the Bayesian network is a conditional probability function. This function provides the conditional distribution of a node v's random variable X_v given the values of the random variables of the node's parents, $\{X_w\colon w \in \mathbf{pa}(v)\}$:

$$\begin{aligned} f_v(\cdot|\cdot)\colon \mathbb{R} \times \mathbb{R}^{|\mathbf{pa}(v)|} &\to \mathbb{R} \\ P(X_v = x_v | X_w = x_w, w \in \mathbf{pa}) &= f_v(x_v | \{x_w\colon w \in \mathbf{pa}(v)\}) \end{aligned}$$

The requirement for these conditional probabilities is the following: for each value x of the random vector X indexed by the nodes of V, we have

$$P(X = x) = \sum_{v \in V} f_v(x_v | \{x_w\colon w \in \mathbf{pa}(v)\}).$$

A Bayesian network is a quadruple (V,E,X,f) satisfying these conditions.

The basic idea of a Bayesian network is a causal structure relating the state variables of interest in the problem. As a very simple example, we consider a problem of medical diagnosis [10]. The directed acyclic graph nodes involve hidden states (the actual condition) and possible opbserved symptoms. The edges carry the conditional probabilities of the symptom given the condition(s). We denote the

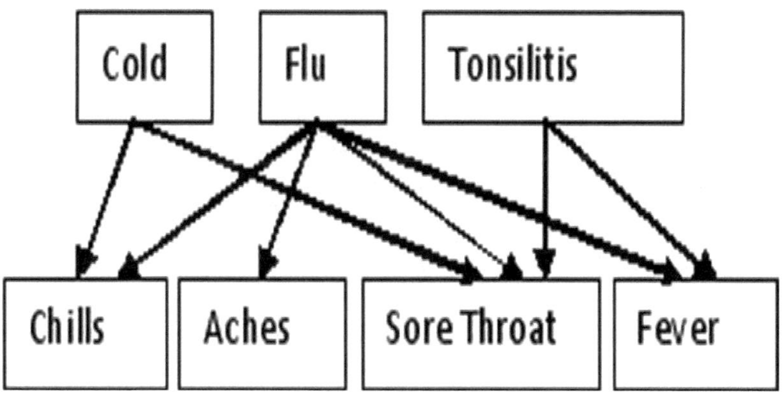

Fig. 2.2 A simple diagnostic network

states by $C_1, C_2, C_3, S_1, S_2, S_3, S_4$ to distinguish the conditions from the symptoms. Each of these is a binary random variable taking a value of 0 for absence and 1 for presence. The joint distribution of conditions and symptoms is formed as in the definition of the Bayesian network. From this we can compute conditionals $S_1, S_2, S_3, S_4 | C_1, C_2, C_3$, and from there Bayes rule allows us to find the posterior of condition likelihoods given symptoms. Maximum a posteriori analysis permits diagnosis.

Additional structure is required for decision theory. In particular, we require a utility function $C = C(u|x)$ valuing controls for different state configurations. The conditional probabilities become

$$f_v(x_v | \{x_w : w \in \mathbf{pa}(v)\}, u)$$

with explicit dependence on the control variable.

More commonly used is the so-called influence diagram [31], whose vertices comprise both state and control variables, as well as values, much like a traditional decision tree. Edges from controls to states model a This modified Bayesian network provides a causal structure to a decision problem.

Consider the problem of UAV perimeter search of Section 2.2. We note that the UAV must choose among three alternatives: move left, move right, or dwell. At any given time, there is potentially a backlog of alerts to be serviced. If the UAV is positioned at an alert site, there is information to be gained by dwelling on that site, but if the UAV is positioned at site without an alert, it would seem natural to move on.

A simple influence diagram for this problem is given in Figure 2.3.

There are significant symmetries in the problem that permit us to view the state of the system in a frame of reference relative to the UAV's position. The quantities relevant to the UAV's decision process are the states at its location, the states to the left, and the states to the right, in terms of alerts and queue lengths. Thus we maintain a state variable of the sites to the left and the sites to the right, along with the current site (which we call "0"). The dynamics shift sites depending on the motion of the UAV, so that if the UAV moves right, the adjacent right position becomes the new 0, and the left-most state shifts to becoming the right-most state due to the closed nature of the discretized perimeter curve.

Recall that the objective criterion proposed and implemented in [8, 20] is given by

$$J(u) = E\Big[\sum_{k=1}^{\infty} \lambda^k (I(x(k), d(k), u(k)) - \beta \max_i(\tau_i(k)),$$

which balances the information gain derived from dwelling against the maximum length of the queues at unvisited alert locations. The parameter *beta* provides a weight that quantifies this balance for the purposes of dynamic programming.

We consider a different weighting, based on the notion that the UAV must choose among three alternatives: dwell, move left, or move right. Seeking to balance the maximum of queue lengths against the information gained at a given site, we choose the following parameterized strategy. If $\Delta I > \gamma \max_i(\tau_i(k))$, that is, if the gain from

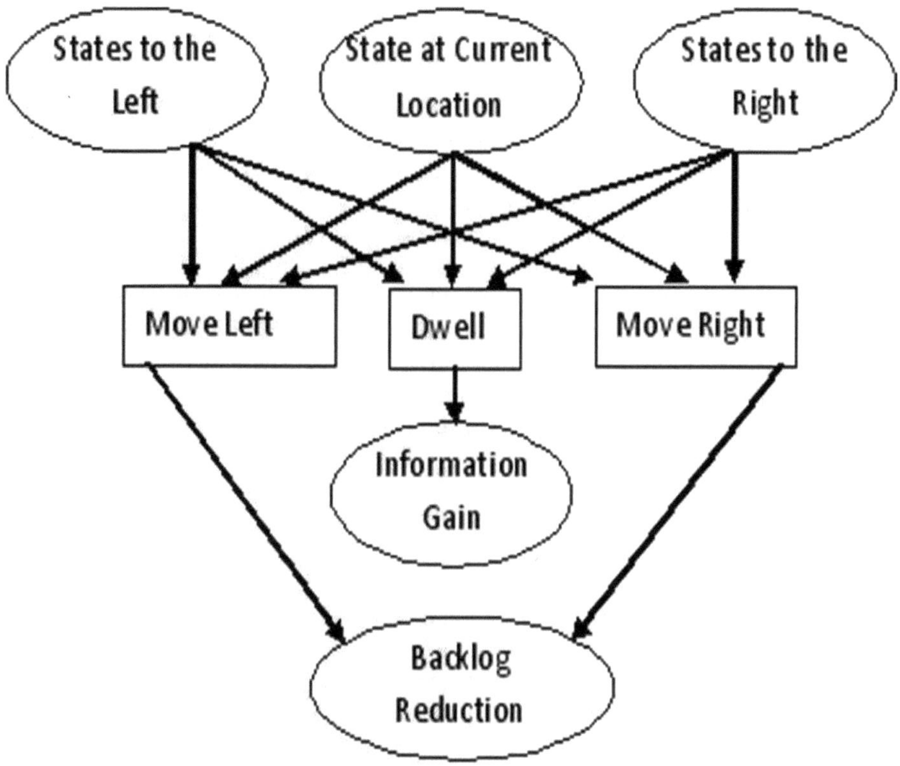

Fig. 2.3 The UAV Decision Influence Diagram

one more dwell is less than (a constant times) the maximum queue's length, then continue to dwell; otherwise, move in the direct of the maximum queue length.

2.5 A Max-Plus Approach to Bayesian Networks and Knowledge Bases

We denote by $\mathbb{R}^e$ the extended real numbers $\mathbb{R}^e = \mathbb{R} \cup \{-\infty\}$. Together with operations $\oplus$ and $\otimes$ defined by

$$a \oplus b = \max\{a, b\},$$
$$a \otimes b = a + b,$$

the set $\mathbb{R}^e$ becomes an idempotent semifield. If Ω is a set with σ-algebra $\mathscr{F}$, we say that $Q\colon \mathscr{F} \to \mathbb{R}^e$ is a max-plus probability measure if

P1 $Q(A) \leq 0$,
P2 $Q(\Omega) = 0$,

P3 $Q(\emptyset) = -\infty$, and
P4 $Q(\cup_{i=1}^{\infty} A_i) = \sup_i Q(A_i)$ for all mutually disjoint $\{A_i\}_{i=1}^{\infty} \subset \mathscr{F}$.

These criteria are completely analogous to the standard definition of probability measures in "plus-times" arithmetic. Note that max-plus probabilities have some unintuitive properties. For example, if B is an event with $P(B) \neq 0$,, then $P(B^c) = 0$, because $0 = P(\Omega) = \max\{P(B), P(B^c)\}$.

We say that $q: \Omega \to \mathbb{R}^e$ is a max-plus probability density for Q if $Q(A) = \sup_{\omega \in A} q(\omega)$ for all $A \in \mathscr{F}$. Densities exist under rather general conditions: if Ω is a second countable topological space, and if $\mathscr{F}$ contains the open sets of Ω, then any max-plus probability measure Q has a unique minimal extension to the power set $\mathscr{P}(\Omega)$ that has a lower semicontinuous max-plus density q [1].

Max plus probability has a close connection with large deviations (see, e.g., [1, 9, 15, 29]). Max-plus algebra can be viewed from a logarithmic point of view. We define the log-plus algebra

$$a \oplus^{\varepsilon} b = \varepsilon \log(e^{a/\varepsilon} + e^{b/\varepsilon}),$$
$$a \otimes^{\varepsilon} b = \varepsilon \log(e^{a/\varepsilon} e^{b/\varepsilon}) = a + b,$$

As noted in [1, 15, 29], the max operation can be obtained as the limit of the log-plus addition operation as $\varepsilon \to 0$. Further, McEneaney and Charalambous [29] note that one can define log-plus probability measures and consider $\varepsilon \to 0$ limits, leading to large deviation results. First, if P is a probability measure in the usual sense, and we define $P^{\varepsilon}(A) = \varepsilon \log(P(A))$, then P^{ε} is a log-plus probability measure, and the limiting measure as $\varepsilon \to 0$ is a max-plus probability measure given by

$$P_0(A) = \begin{cases} 0 & \text{if } P(A) > 0 \\ -\infty & \text{if } P(A) = 0 \end{cases}$$

Next, we note that if $(\Omega, \mathscr{F}, P^{\varepsilon})$ is a family of complete probability spaces satisfying a large deviation principle [37], then the limiting measure $\mu(A) = \lim_{\varepsilon \to 0} P^{\varepsilon}(A)$ is a max-plus probability measure. Recall that the large deviation principle requires the following conditions to hold: There exists an upper semicontinuous function $I: \Omega \to [-\infty, 0]$ satisfying

LD1 For each $m > -\infty$, the set $\{x \in \Omega : m \leq I(x)\}$ is compact,
LD2 For each closed set $C \in \mathscr{F}$, we have that

$$\lim_{\varepsilon \to 0} \sup \varepsilon \log P^{\varepsilon}(C) \leq \sup_{x \in C} I(x),$$

LD3 For each open set $O \in \mathscr{F}$, we have that

$$\lim_{\varepsilon \to 0} \inf \varepsilon \log P^{\varepsilon}(O) \geq \sup_{x \in O} I(x).$$

The function I is the max-plus density of the max-plus probability measure μ.

Entropy and information [9] can also be defined in terms of max-plus measures and large deviations. Indeed, the entropy is just the negative of the max-plus probability, which literally becomes the information.

We say that X is a max-plus random variable if it is a measureable mapping from Ω to $\mathbb{R}^e$. Expectations in the max-plus sense are computed as follows:

$$\mathbb{E}(X) = \sup_{\omega}\{X(\omega) + q(\omega)\}$$

when the measure Q has density q. More generally, one may construct max-plus versions of the Lebesgue integral for expectation, but we will not need that level of generality here.

We also consider max-plus conditional probabilities: $Q(A|B) = Q(A) - Q(A \cap B)$.

The max-plus version of Bayes' theorem is as follows. Given a measureable partition $H_i, i = 1, 2, \cdots, \infty$ and a set $A \in \mathscr{F}$, we have

$$Q(H_i|A) = Q(A|H_i) + Q(H_i) - \sup_{j=1}^{\infty}\{Q(A|H_j) + Q(H_j)\}.$$

With conditional probabilities in hand, we may proceed to max-plus Bayesian networks. Given a max-plus probability space $(\Omega, \mathscr{F}, Q)$ and a directed acyclic graph (V, E), we require a set of max-plus random variables $X_v, v \in V$ and a function $f\colon V \times \mathbf{pa}(V) \to \mathbb{R}^e$ satisfying

$$Q(X = x) = \sup_{v \in V} f(x_v|\mathbf{pa}(v))$$

for all $x \in \mathbb{R}^e$.

We interpret the max-plus conditional probabilities $f(x_v|\mathbf{pa}(v))$ as the cost of state x_v for node v given the parent node states $x_w, w \in \mathbf{pa}(v))$.

Reconsidering the example problem of Section 2.2, we pose the problem in terms of a linear information gain in the max-plus probability sense:

$$I(d) = \max\{P_C^0 T(d) + p, P_C^0 N(d)\},$$

in which

$$\begin{aligned} P_{CT}(d) &= A_T + B_T d \\ P_{CN}(d) &= A_N + B_N d, \end{aligned}$$

provide the max-plus probabilities of correct identification of target and nuisance, respectively. We use linear functions motivated by the large deviation interpretation. Applying large deviation reasoning to the alert arrival process, we have the max-plus probability of waiting at least t for an arrival as $-\lambda t$.

The max-plus interpretation of the Bayesian network and influence diagram leads us to consider a decision rule of same form as the . If $\Delta I > \gamma \max_i(\tau_i(k))$, that is, if the gain from one more dwell is less than (a constant times) the maximum queue's length, then continue to dwell; otherwise, move in the direct of the maximum queue length.

Acknowledgements. The author gratefully acknowledges the support of the Air Force Office of Scientific Research grants FA9550-09-1-0524 and FA9550-10-1-0499.

References

1. Akian, M.: Densities of idempotent measures and large deviations. Trans. AMS 351(11), 4515–4543 (1999)
2. Akian, M., Quadrat, J.P., Viot, M.: Duality between probability and optimization. In: Gunawardena, J. (ed.) Idempotency. Cambridge University Press, Cambridge (1998)
3. Behrens, T., Woolrich, M., Walton, M., Rushworth, M.: Learning the value of information in an uncertain world. Nature Neurosci. 10(9), 1214–1221 (2007)
4. Berger, J.: Statistical Decision Theory and Bayesian Analysis. Springer, New York (1985)
5. Bladon, P., Day, P.S., Hughes, T., Stanley, P.: High-Level Fusion using Bayesian Networks: Applications in Command and Control. In: Information Fusion for Command Support: Meeting Proceedings RTO-MP-IST-055, Paper 4, pp. 4-1–4-18. RTO, Neuilly-sur-Seine (2006)
6. Bourgault, F., Ahmed, N., Shah, D., Campbell, M.: Probabilistic Operator-Multiple Robot Modeling Using Bayesian Network Representation. In: Proc AIAA GNC, Paper AIAA 2007-6589 (2007)
7. Buxton, H.: Advanced Visual Surveillance Using Bayesian Networks. In: IEE Colloquium on Image Processing for Security Applications (Digest No.: 1997/074), pp. 9/1–9/5 (1997)
8. Chandler, P., Hansen, J., Holsapple, R., Darbha, S., Pachter, M.: Optimal Perimeter Patrol Alert Servicing with Poisson Arrival Rate. In: Proc. AIAA GNC, Chicago, IL (2009)
9. Charalambous, C.D., Djouadi, S.M.: Large Deviations and Deterministic Measures of Information. In: Proc. European Control Conf. (September 2003)
10. Darwiche, A.: Modeling and Reasoning with Bayesian Networks. Cambridge University Press, Cambridge (2009)
11. Donley, M., Schwartz, N.: United States Air Force Unmanned Aircraft Systems Flight Plan, pp. 2009–2047. Headquarters, United States Air Force, Washington DC (2009)
12. Donley, M., Schwartz, N.: Report on Technology Horizons: A Vision for Air Force Science and Technology During 2010-2030. Technical Report AF/ST-TR-10-01-PR, United States Air Force, Washington DC (2010)
13. El-Gamal, M., Grether, D.: Are People Bayesian? Uncovering Behavioral Strategies. JASA 90(432), 1137–1145 (1995)
14. Fennema, H., Wakker, P.: Original and Cumulative Prospect Theory: A Discussion of Empirical Differences. J. Behavioral Dec. Making 10, 53–64 (1997)
15. Fleming, W., Kaise, H., Sheu, S.-J.: Max-Plus Stochastic Control and Risk-Sensitivity. Appl. Math. Opt. 62(1), 81–144 (2010)
16. Griffiths, T., Tenenbaum, J.: Optimal Predictions in Everyday Cognition. Psych. Sci. 17(9), 767–773 (2006)
17. Han, S.H., McEneaney, W.M.: Solution of an optimal sensing and interception problem using idempotent methods. In: Fahroo, F., Wang, L.Y., Yin, G. (eds.) Recent Adv. in Res. on Unmanned Aerial Veh. LNCIS, vol. 444, pp. 47–67. Springer, Heidelberg (2013)
18. Kahneman, D.: A Perspective on Judgment and Choice: Mapping Bounded Rationality. Amer. Psych. 58(9), 697–720 (2003)
19. Kahneman, D., Tversky, A.: Prospect Theory: an Analysis of Decision under Risk. Econometrica 47(2), 263–291 (1979)

20. Kalyanam, K., Pachter, M., Darbha, S., Chandler, P.: Approximate Value Iteration with State Aggregation Applied to Perimeter Patrol. Intl. J. Robust Nonlinear Control 21(12), 1396–1409 (2011)
21. Kessel, R.: The Burden of Computer Advice: Using Expert Systems as Decision Aids. DRDC Atlantic Technical Report TR 2003-241, Defence R & D Canada (2003)
22. Klein, G.: Sources of Power: How People Make Decisions. MIT Press, Cambridge (1999)
23. Koski, T., Noble, J.: Bayesian Networks: An Introduction. Wiley, Chichester (2009)
24. Levitt, T., Leister, K., Woodaman, R., Askvig, J., Laskey, K., Hayes-Roth, R.: Valuing Persistent ISR Assets, `http://faculty.nps.edu/fahayesr/RapidProVIRT.html`
25. Lipshitz, R., Klein, G., Orasanu, J., Salas, E.: Taking Stock of Naturalistic Decision Making. J. Behavioral Dec. Making 14, 331–352 (2001)
26. Linegang, M., Haimson, C., MacMillan, J., Freeman, J.: Human Control in Mixed-Initiative Systems: Lessons from the MICA-SHARC Program. In: Proc. IEEE Systems, Man, Cyber. (2003)
27. Lipshitzand, R., Strauss, O.: Coping with Uncertainty: a Naturalistic Decision Making Analysis. Org. Behavior and Human Dec. Proc. 69, 149–163 (1997)
28. Martins, A.: Probability Biases as Bayesian Inference. Judg. and Dec. Making 1(2), 108–117 (2006)
29. McEneaney, W.M., Charalambous, C.D.: Large Deviations Theory, Induced Log-Plus and Max-Plus Measures and their Applications. In: Proc. Mathematical Theory of Network and Systems 2000, Perpignan, France, June 19-23 (2000)
30. Osinga, F.: Science, strategy, and war: The strategic theory of John Boyd. Routledge, New York (2007)
31. Pearl, J.: Probabilistic Reasoning in Intelligent Systems: Networks of Plausible Inference. Morgan Kaufmann, San Mateo (1988)
32. Peng, X., Xu, D.: Intelligent Online Path Planning for UAVs in Adversarial Environments. Int. J. Adv. Robotic Sy. 9, 1–12 (2012)
33. Peng, X., Xu, D., Yan, W.: Intelligent Flight for UAV via Integration of Dynamic MOEA, Bayesian Network and Fuzzy Logic. In: Proc. 50th IEEE CDC, pp. 3870–3875 (2011)
34. Quadrat, J.P.: Max-Plus Working Group, "Min-plus linearity and Statistical Mechanics". Markov Processes and Related Fields 3(4), 565–587 (1998)
35. Santos Jr., E., Santos, E.S.: A framework for building knowledge-bases under uncertainty. J. Exp. Theor. Artif. Intell. 11(2), 265–286 (1999)
36. Simon, H.: A Behavioral Model of Rational Choice. In: Models of Man, Social and Rational: Mathematical Essays on Rational Human Behavior in a Social Setting. Wiley, New York (1957)
37. Stroock, D.W.: An Introduction to the Theory of Large Deviations. Springer, New York (1984)
38. Trommershauser, J., Landy, M., Maloney, L.: Humans Rapidly Estimate Expected Gain in Movement Planning. Psych. Sci. 17(11), 981 (2006)
39. Tversky, A., Kahneman, D.: Advances in Prospect Theory: Cumulative Representation of Uncertainty. J. Risk Uncertainty 5, 297–323 (1992)
40. Woods, D.: Decomposing automation: Apparent simplicity, real complexity. In: Parasuraman, R., Mouloua, M. (eds.) Automation and Human Performance: Theory and Applications, pp. 1–16. Erlbaum, Mahwah (1996)

Chapter 3
Solution of an Optimal Sensing and Interception Problem Using Idempotent Methods

Seung Hak Han and William M. McEneaney

Abstract. We consider a problem where one desires to intercept a vehicle before it reaches a specific (target) location. The vehicle travels on a road network. There are unattended ground sensors (UGSs) on the road network. The data obtained by a UGS is uploaded to an unmanned aerial vehicle (UAV) when the UAV overflies the UGS. There is an interceptor vehicle which may travel the road network. The interceptor uses intruder state estimates based on the data uploaded from the UGSs to the UAV. There is a negative payoff if the intruder is intercepted prior to reaching the target location, and a positive payoff if the intruder reaches the target location without having been previously intercepted. The intruder position is modeled as a Markov chain. The observations are corrupted by random noise. A dynamic programming approach is taken where the value is propagated using a min-plus curse-of-dimensionality-free algorithm. The double-description method for convex polytopes is used to reduce computational loads.

3.1 Introduction

We consider a problem in coordinated command and control. The problem belongs to a class of problems where the state of the system consists not only of the physical state, but also a component describing the information state [9]. Although the payoff will depend entirely on the physical state, achievement of the desired physical-state goal depends on the controls used to advance the information state process. There are controls which affect

Seung Hak Han · William M. McEneaney
University of California San Diego, Dept. Mech. and Aero. Eng., La Jolla, CA 92093-0411, USA
e-mail: {shh013,wmceneaney}@ucsd.edu

F. Fahroo et al. (Eds.): *Recent Adv. in Res. on Unmanned Aerial Veh.*, LNCIS 444, pp. 47–67.
DOI: 10.1007/978-3-642-37694-8_3

both the physical- and information-state processes, and controls which only affect the latter. One difficult aspect of these problems is that the information state is typically a high-dimensional object. In the problem studied herein, even though the physical state will take values in a finite set, the information state will be modeled in terms of a probability distribution over that finite set, and lies in a Euclidean space where the dimension is the cardinality of the finite set. Consequently, one needs computational tools which can handle control problems over high-dimensional spaces. Natural approaches are the algorithms in the class of idempotent curse-of-dimensionality-free methods [5, 7, 8]. Although such methods have been applied to similar information state problems before [9, 13], here we combine those techniques with the "double-description method" [12], which will prove to be a fruitful combination in terms of computational-effort reduction.

The specific application under consideration is as follows, where we note that it may be helpful to refer to Figure 3.1. There is a single intruder vehicle. We would like to intercept that vehicle before it reaches a specific target location, labeled as the base in the figure. The intruder may only move along the given road network. There are unattended ground sensors (UGSs) placed along the road network. The intruder is unaware of the locations of these sensors. When the intruder passes through a UGS location, the data is acquired and stored by the UGS. We have an unmanned aerial vehicle (UAV) which is monitoring the situation. When the UAV overflies a USG, the data from the UGS is uploaded to the UAV. We also have an interceptor ground vehicle which moves along the road network. We will use a discrete-time, discrete-state model of the physical system state. The problem has an exit-time aspect. The exit time is the first time when either the interceptor and the intruder are co-located or the intruder reaches the base. We will be attempting to minimize the payoff. There will be a positive payoff if the intruder reaches the base, and a negative payoff if the interceptor is co-located with the intruder prior to the intruder reaching the base.

We will employ stochastic models. Purely game-theoretic and more general stochastic game models could certainly be considered. The reason for applying a stochastic approach over a game-theoretic approach is that we expect significant observation errors, and that these would be best modeled as random processes. The more general stochastic game approach might be optimal, but the computational burden is already significant. One possible alternative which might be computationally tractable is a risk-averse stochastic model, as these can prove effective in the context of stochastic games [10].

With a stochastic control problem formulation, we will have an information state which takes the form of an observation-conditioned probability

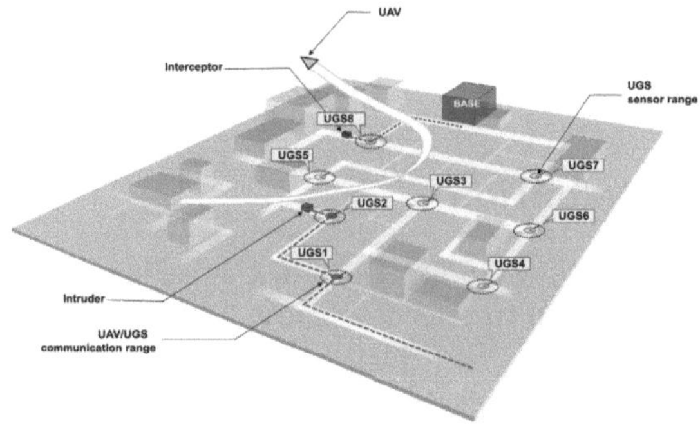

Fig. 3.1 UAV/UGS/Interceptor problem

distribution over the space of possible intruder locations. As we assume full communication between the UAV, the interceptor and possibly a central processing location, we do not distinguish where the computations may be occurring. Updates to the conditional distribution will be through Bayes' rule when observations are processed. As the state-space is finite, we model the intruder movement as a Markov chain. The state of the system at any given moment has both a physical and an informational component. Of course the informational component is the conditional distribution over a set of locations. The physical component will consist of the locations of the intruder, the UAV and the interceptor. Our controls at each time step will be the next UAV location and the next interceptor location. We will assume that the UAV can move from any location to any other location in one step, whereas in one step, both the interceptor and the intruder are constrained to move only between adjacent nodes of the road-network graph.

As our main focus will be on the application of idempotent curse-of-dimensionality-free methods to this class of control problems, we will make some other simplifications so as to reduce the technical complexity. Most notable is the following. Each UGS takes data each time the intruder passes by. However, our control system only becomes aware of the observations taken by a UGS when the UAV overflies that UGS. For example, at time t_5, we may obtain data from UGS 7, where it observed the intruder at time t_4. Then, at time t_6, we might obtain data from UGS 3, where it observed the intruder at time t_2. If we have already processed the time t_4 data, updating the system with the time t_2 data becomes a bit tedious. One could maintain the full history of the conditional probability distributions so that one could go back and fold in the t_2 data for time t_2 and then propagate forward back up to time t_4 to then process that data, and further propagate forward to the current time. Although the requisite algorithm would be complex, it would not substantially affect the total computational cost, which is the central point of the discussion here. Consequently, we restrict ourselves to models where we only process newly uploaded observations which are more recent than the most recent previously uploaded observation. A comprehensive discussion of the issues related to the sensor data timing can be found in [3].

3.2 Problem Definition

Assume that there are N_s UGS locations in the region under consideration. (As above, it may be useful to refer to Figure 3.1.) The set of these locations may be indexed as $\mathcal{N}_s \doteq]1, N_s[\doteq \{1, 2, \cdots, N_s\}$. Throughout, for integers $a \leq b$, we will use $]a, b[$ to denote $\{a, a+1, a+2, \ldots b\}$. We denote the target, or base, location as $N_b = N_s + 1$ and let $\mathcal{N}_b \doteq]1, N_b[=]1, N_s + 1[$. We will let the states of the intruder and the interceptor take values in the finite set $\mathcal{N}_b$, with the addition of one more state value to be indicated just below. The road network will be interpreted as a graph with the set of nodes being $\mathcal{N}_b$, and an edge, (i, j) (equivalently (j, i)), existing between nodes i and j if there exist roads connecting those nodes without passing through another UGS location. We let the neighborhood, or set of immediately accessible states, from node $i \in \mathcal{N}_b$, be the set of nodes, $j \in \mathcal{N}_b$ such that there exists an edge (i, j) connecting them. and denote this set as $\mathcal{I}^i$. Recall that we use a discrete-time model. For simplicity, we let the time-step be one, and let the time be denoted by $t \in \mathcal{T} \doteq \{0, 1, 2, \ldots\}$. We can model the intruder motion process as a Markov chain on $\mathcal{N}_b$, with transition probability matrix, $I\!\!P \in I\!\!R^{N_b \times N_b}$. In particular, we will take N_b to be an absorbing state.

The interceptor also moves on $\mathcal{N}_b$. However, in this case, we will have an interceptor control which will allow the interceptor to move from $i \in \mathcal{N}_b$ to any $j \in \mathcal{I}^i$. Recall that the intruder is "captured" if the interceptor and intruder are co-located. Because of this, we find it helpful to add another location to the model. We suppose that if the interceptor and intruder are co-located at time, t, then the intruder moves instantaneously with probability one to an exit-state location denoted by N_e. Therefore, we add this location to $\mathcal{N}_b$, and let $\mathcal{N}_e \doteq]1, N_e[=]1, N_s + 2[$.

The intruder and interceptor states at time $t \in \mathcal{T}$ are denoted by $x_t^R \in \mathcal{N}_e$ and $x_t^i \in \mathcal{N}_b$, respectively. The interceptor control at time t, is denoted by $u_t^i \in \mathcal{I}^{x_t^i} \subseteq \mathcal{N}_b$. The interceptor dynamics are simply $x_{t+1}^i = u_t^i$. The UAV state at time t is denoted by x_t^o. As the UAV may travel from any node to any other in one time step, we let the UAV control at time t be $u_t^o \in \mathcal{N}_b$, and the UAV dynamics are simply $x_{t+1}^o = u_t^o$.

Our information at time t on the intruder states in $\mathcal{N}_e$ is described by probability distribution q_t. In particular, the jth component of q_t, q_t^j, is the probability that the intruder is at $j \in \mathcal{N}_e$ at time t. Distribution q_t lies in probability simplex S^{N_e} where

$$S^{N_e} \doteq \left\{ q \in I\!\!R^{N_e} \,|\, q^j \in [0,1] \ \forall j \in \mathcal{N}_e \text{ and } \sum_{j \in \mathcal{N}_e} q^j = 1 \right\}.$$

3.2.1 System Dynamics

At this point, we have indicated some components of the problem statement. First, note that the system state at time t will be $(x_t^o, x_t^i, x_t^R, q_t)$, and the controls will be (u_t^o, u_t^i). We have the dynamics for x^o and x^i, which are

$$x_{t+1}^o = u_t^o \quad \text{and} \quad x_{t+1}^i = u_t^i, \tag{3.1}$$

where

$$u_t^o \in \mathcal{N}_s \quad \text{and} \quad u_t^i \in \mathcal{I}^{x_t^i} \subseteq \mathcal{N}_b.$$

We have indicated that $x_\cdot^R$ propagates as a Markov chain, with nominal transition matrix $I\!P \in I\!R^{N_b \times N_b}$. However, because of the potential for interception, we now extend and modify that transition matrix. We suppose that some interceptor control, u_t^i is given. For any $u^i \in \mathcal{N}_b$, we let $\widehat{I\!P}^{u^i} \in I\!R^{N_e \times N_e}$ be as follows.

$$\widehat{I\!P}^{u^i}_{j,k} \doteq \begin{cases} I\!P_{j,k} & \text{if } j,k < N_b,\ k \neq u^i \\ 0 & \text{if } j,k < N_b,\ k = u^i \\ I\!P_{j,k} & \text{if } j < N_b,\ k = N_b \\ I\!P_{j,u^i} & \text{if } j < N_b,\ k = N_e \\ 0 & \text{if } j = N_e, k \neq N_e \text{ or } j = N_b, k \neq N_b \\ 1 & \text{if } j = N_e, k = N_e \text{ or } j = N_b, k = N_b. \end{cases}$$

Lastly, we need to describe the dynamics of the information state, q. We suppose that at each time step, the observation update occurs prior to the dynamics update, where the latter is described by the above transition matrix, $\widehat{I\!P}^{u^i}$. We now describe the observation update, where this will be performed via Bayes' rule.

Each UGS may record information regarding the intruder when the intruder is in its range of detection, such as the time, position, speed, direction, and type of intruder. As our focus is on the development of an efficient numerical algorithm, we will use a simple UGS observation model. We suppose that if the UAV control at time t is $u_t^o \in \mathcal{N}_s$, then the UAV immediately uploads the (random variable) observation at time t, y_t, made by the UGS. We let y_t take values in $\mathcal{Y} \doteq \{0,1\}$, where $y_t = 0$ is the observation that no intruder passed the node $j = u_t^o$ at time t and $y_t = 1$ is the observation that the intruder passed there at that time. Again, for a more complete analysis of observation processing where multiple time-step information may be uploaded to the UAV, see [3].

We suppose that the UGSs have known false alarm and misdetection rates, and that there exists noise in the communication system between the UAV and the UGSs, also with known parameters, although communication dropout is not considered. With this, we obtain an overall confusion matrix. In particular, we let $R_j^{u^o,y}$ denote the probability of observation $y \in \mathcal{Y}$ given the UAV is at $u^o \in \mathcal{N}_s$ and the intruder is at $j \in \mathcal{N}_e$. That is,

$$R_j^{u^o,y} \doteq P_{u^o}(y_t = y \,|\, x_t^R = j), \tag{3.2}$$

where the subscript u^o indicates the dependence of the particular UAV sensing control. Let $\mathbf{R}^{u^o,y}$ be the vector of length N_e with components $R_j^{u^o,y}$. Throughout, for generic vector, $\mathbf{R}$, we let $D(\mathbf{R})$ denote the diagonal matrix with diagonal entries given by the components of $\mathbf{R}$. More specifically, $D(\mathbf{R}^{u^o,y})$ denotes the $N_e \times N_e$ matrix with components $[D(\mathbf{R}^{u^o,y})]_{jj} = R_j^{u^o,y}$, and such that $[D(\mathbf{R}^{u^o,y})]_{ij} = 0$ for $i \neq j$.

Suppose that under control u_t^o, the observation $y_t = y \in \mathcal{Y}$ is made. Then, by Bayes' rule, the a posteriori probability is given by

$$\hat{q}_t^j \doteq P_{u_t^o}(x_t^R = j \,|\, y_t = y) = \frac{P_{u_t^o}(y_t = y \,|\, x_t^R = j) P(x_t^R = j)}{\sum_{j \in N_e} P_{u_t^o}(y_t = y, x_t^R = j)}$$

$$= \frac{P_{u_t^o}(y_t = y \,|\, x_t^R = j) P(x_t^R = j)}{\sum_{j \in N_e} P_{u_t^o}(y_t = y \,|\, x_t^R = j) P(x_t^R = j)},$$

which by (3.2) and the definition of $\mathbf{R}^{u_t^o,y_t}$,

$$= \frac{R_j^{u_t^o,y} q_t^j}{\mathbf{R}^{u_t^o,y} \cdot q_t}.$$

Hence, by the definition of $D(\cdot)$,

$$\hat{q}_t = \frac{1}{\mathbf{R}^{u_t^o,y} \cdot q_t} D(\mathbf{R}^{u_t^o,y}) q_t \doteq \hat{\beta}^{u_t^o,y}(q_t) \tag{3.3}$$

where $\hat{q}_t \in S^{N_e}$.

Consequently, given the current interceptor state x_t^i and the interceptor control $u_t^i \in \mathcal{I}^{x_t^i}$, the combined dynamics and observation update becomes

$$q_{t+1} = \left(\widehat{I\!P}^{u_t^i}\right)^T \hat{\beta}^{u_t^o,y}(q_t). \tag{3.4}$$

We remark that there is an implicit observation by the interceptor as well as the UAV observation discussed above. For example, if there is no intruder at $x_{t+1}^i = u_t^i$, then there is an implicit interceptor-observation that the intruder is not at this location. One should assume that this observation is noise-free as we also assume the intruder is captured with probability one if they are co-located. In this case, letting $y = 0$ denote the observation that the intruder is not at the interceptor location, one is led to an $R_j^{u_t^i,y} = R_j^{u_t^i,0}$ (analogous to $R_j^{u_t^o,y}$ above) which is one if $j \neq u_t^i$ and zero otherwise. Because of the form of $\widehat{P}^{u^i}$ above, including these additional observation dynamics into the backward dynamic program below yields absolutely no difference in the dynamic programming update. (Of course, in forward simulation, one does need to include this implicit observation.)

3.2.2 Payoff and Value

We impose a finite time-horizon, T, on the problem. We will not start from basic principles in definition of the payoff and value function. Proceeding from that point to the problem definition we use below would involve a substantial argument using the Separation Principle, which is outside the scope of our interests here. Let $\mathcal{T}_T \doteq]0,T[$ and $\mathcal{T}_T^- \doteq]0,T-1[$. For $t \in \mathcal{T}_T^-$, let

$$\begin{aligned}\mathcal{U}_t^i \doteq \Big\{ \{u_r^i\}_{r=t}^{T-1} : [\mathcal{N}_b \times S^{N_e}]^{T-t} \to [\mathcal{N}_b]^{T-t} \;\Big|\; & \text{if } (x^i, q), (\tilde{x}^i, \tilde{q}) \in [\mathcal{N}_b \times S^{N_e}]^{T-t} \\ & \text{and } r \in]t, T-1[\text{ are such that } (x^i, q)_{]t,r[} = (\tilde{x}^i, \tilde{q})_{]t,r[}, \\ & \text{then } u_r^i(x_\cdot^i, q_\cdot) = u_r^i(\tilde{x}_\cdot^i, \tilde{q}_\cdot) \Big\}.\end{aligned}$$

We define $\mathcal{U}_t^o$ similarly, but with $\mathcal{N}_b$ in the range replaced by $\mathcal{N}_s$. Let $\bar{\beta} \in \mathbb{R}^{N_e}$ where $\bar{\beta}_{N_b} = c_b$, $\bar{\beta}_{N_e} = c_e$, and $\bar{\beta}_i = 0$ otherwise. Here, c_b denotes the payoff obtained in the case that the intruder reaches the base, and c_e denotes the payoff in the case that the intruder is intercepted prior to reaching the base. Let $\psi : S^{N_e} \to \mathbb{R}$ be given by

$$\psi(q) = \bar{\beta} \cdot q.$$

For $t \in \mathcal{T}_T^-$, let $\tilde{\mathcal{Y}}_t$ denote the observation process $y_{]t,T-1[}$, taking values in $\mathcal{Y}$, where we suppose the $y_t | x_t^R$ form a sequence of independent random variables. For $t \in \mathcal{T}_T$, the payoff $\mathcal{J}_t : S^{N_e} \times \mathcal{N}_e \times \mathcal{U}_t^i \times \mathcal{U}_t^o \to \mathbb{R}$ is given by

$$\mathcal{J}_t(q, x^i, u_\cdot^i, u_\cdot^o) \doteq \mathbf{E}_{y_{]t,T-1[}} \psi(q_T),$$

where $q_\cdot$ satisfies (3.4) with $q_t = q$, and $x_\cdot^o, x_\cdot^i$ satisfy (3.1) with $x_t^i = x^i$. For $t \in \mathcal{T}_T$, the value $\overline{V}_t : S^{N_e} \times \mathcal{N}_e \to \mathbb{R}$ is given by

$$\overline{V}_t(q, x^i) \doteq \inf_{u_\cdot^i \in \mathcal{U}_t^i} \inf_{u_\cdot^o \in \mathcal{U}_t^o} \mathcal{J}_t(q, x^i, u_\cdot^i, u_\cdot^o). \tag{3.5}$$

3.3 Idempotent Based Dynamic Programming

To solve the state feedback problem, backward dynamic programming will be enabled via an idempotent-based algorithm, not subject to the curse-of-dimensionality. As it is standard given the value function definition (3.5), we present the dynamic programming principle without proof. For $t \in \mathcal{T}_T^-$ we have

$$\overline{V}_t(q, x^i) = \min_{u^o \in \mathcal{N}_s} \min_{u^i \in \mathcal{I}^{x^i}} \mathbf{E}_{y_t} \left\{ \overline{V}_{t+1} \left[\left(\widehat{\mathbb{P}}^{u^i} \right)^T \hat{\beta}^{u^o, y_t}(q), u^i \right] \right\}, \tag{3.6}$$

where the expectation is over the set of possible observations at time t.

For any set $\mathcal{Z}$ and positive integer M, let $|\mathcal{Z}|$ be the cardinality of $\mathcal{Z}$ and $\mathcal{P}^M(\mathcal{Z})$ be the set of all sequence of length M with elements from $\mathcal{Z}$. Let the cardinality of the observation set $\mathcal{Y}$ be denoted by N_y, and given x^i, $N_u^{x^i} \doteq |\mathcal{N}_s \times \mathcal{I}^{x^i}|$. Given x^i and time t, we denote the index set of coefficient vectors of functionals by $\mathcal{Z}_t^{x^i} \doteq]1, |\mathcal{Z}_t^{x^i}|[$. It will be helpful to define $\mathcal{G} : I\!R^{N_s} \to I\!R^{N_e}$ given by

$$\mathcal{G}(b) \doteq (b^T, c_b, c_e)^T,$$

and $\mathcal{G}^{-1} : \text{Range}(\mathcal{G}) \to I\!R^{N_s}$ given by

$$\mathcal{G}^{-1}(\bar{b}) = \bar{b}_{]1,N_s[},$$

where $\bar{b}_{]1,N_s[}$ denotes the vector consisting of the first N_s components of $\bar{b}$.

Lemma 1. *Suppose there exist index sets $\mathcal{Z}_{t+1}^x$ (of cardinality Z_{t+1}^x) for all $x \in \mathcal{N}_b$, and $b_{t+1}^{x,z} \in I\!R^{N_s}$ for all $x \in \mathcal{N}_b$ and all $z \in \mathcal{Z}_{t+1}^x$ such that*

$$\overline{V}_{t+1}(q, x^i) = \min_{z \in \mathcal{Z}_{t+1}^{x^i}} \bar{b}_{t+1}^{x^i,z} \cdot q, \quad \forall q \in S^{N_e},\ \forall x^i \in \mathcal{N}_b,$$

where $\bar{b}_{t+1}^{x^i,z} = \mathcal{G}(b_{t+1}^{x^i,z})$. Then,

$$\overline{V}_t(q, x^i) = \left\{ \min_{u^o \in \mathcal{N}_s} \min_{u^i \in \mathcal{I}^{x^i}} \sum_{y \in \mathcal{Y}} \min_{z \in \mathcal{Z}_{t+1}^{u^i}} \left[(D(\tilde{\mathbf{R}}^{u^o,y}) \widetilde{I\!P}^{u^i} \bar{b}_{t+1}^{u^i,z})^T \mathcal{H}(q) \right] \right\}$$
$$+ c_b q^{N_b} + c_e q^{N_e},$$

where $\widetilde{I\!P}^{u^i}$ denotes the first N_s rows of $\widehat{I\!P}^{u^i}$, $\tilde{\mathbf{R}}^{u^o,y}$ denotes the first N_s elements of $\mathbf{R}^{u^o,y}$, and $\mathcal{H} : S^{N_e} \to I\!R^{N_s}$ be given by $\mathcal{H}(q) = q^{]1,N_s[}$ (i.e., the first N_s elements of q for any $q \in S^{N_e}$).

Proof. First, fix any $u_t^o = u^o \in \mathcal{N}_b$, and let y_t be the resulting observation. Suppose x_t^R is distributed according to q. We have

$$\begin{aligned} P_{u^o}(y_t = y) &= \sum_{i \in \mathcal{N}_e} P_{u^o}(y_t = y, x_t^R = i) \\ &= \sum_{i \in \mathcal{N}_e} P_{u^o}(y_t = y \,|\, x_t^R = i) P(x_t^R = i) \\ &= \sum_{i \in \mathcal{N}_e} R_i^{u^o,y} q^i = \mathbf{R}^{u^o,y} \cdot q. \end{aligned} \tag{3.7}$$

Now recall from (3.6) that

$$\overline{V}_t(q, x^i) = \min_{u^o \in \mathcal{N}_s} \min_{u^i \in \mathcal{I}^{x^i}} \mathbf{E}_{y_t} \left\{ \overline{V}_{t+1} \left[\left(\widehat{I\!P}^{u^i} \right)^T \hat{\beta}^{u^o,y}(q), u^i \right] \right\},$$

which with the assumed form,

$$= \min_{u^o \in \mathcal{N}_s} \min_{u^i \in \mathcal{I}^{x^i}} \mathbf{E}_{y_t} \left\{ \min_{z \in \mathcal{Z}^{u^i}_{t+1}} \bar{b}^{u^i,z}_{t+1} \cdot \left(\widehat{I\!P}^{u^i} \right)^T \hat{\beta}^{u^o,y}(q) \right\}$$

$$= \min_{u^o \in \mathcal{N}_s} \min_{u^i \in \mathcal{I}^{x^i}} \sum_{y \in \mathcal{Y}} \left\{ \min_{z \in \mathcal{Z}^{u^i}_{t+1}} \left[\bar{b}^{u^i,z}_{t+1} \cdot \left(\widehat{I\!P}^{u^i} \right)^T \hat{\beta}^{u^o,y}(q) \right] P_{u^o}(y_t = y) \right\}.$$

Using (3.3) and (3.7), this becomes

$$\overline{V}_t(q, x^i) = \min_{u^o \in \mathcal{N}_s} \min_{u^i \in \mathcal{I}^{x^i}} \sum_{y \in \mathcal{Y}} \left\{ \min_{z \in \mathcal{Z}^{u^i}_{t+1}} \left(\bar{b}^{u^i,z}_{t+1} \right)^T \right.$$

$$\left. \left[\left(\widehat{I\!P}^{u^i} \right)^T \frac{1}{\mathbf{R}^{u^o,y} \cdot q} D(\mathbf{R}^{u^o,y}) q \right] \mathbf{R}^{u^o,y} \cdot q \right\}$$

$$= \min_{u^o \in \mathcal{N}_s} \min_{u^i \in \mathcal{I}^{x^i}} \sum_{y \in \mathcal{Y}} \min_{z \in \mathcal{Z}^{u^i}_{t+1}} \left(\bar{b}^{u^i,z}_{t+1} \right)^T \left(\widehat{I\!P}^{u^i} \right)^T D(\mathbf{R}^{u^o,y}) q. \tag{3.8}$$

Now, for $u^i \in \mathcal{N}_b$, let $\widetilde{I\!P}^{u^i}$ denote the first N_s rows of $\widehat{I\!P}^{u^i}$. Note that $\widehat{I\!P}^{u^i}_{N_b,j} = \delta_{N_b,j}$ and $\widehat{I\!P}^{u^i}_{N_e,j} = \delta_{N_e,j}$, where $\delta_{i,j}$ denotes the Dirac delta function. Then, for any $q \in S^{N_e}$,

$$\left(\widehat{I\!P}^{u^i} \right)^T D(\mathbf{R}^{u^o,y}) q = \left(\widetilde{I\!P}^{u^i} \right)^T \mathcal{H}(D(\mathbf{R}^{u^o,y}) q) + q', \tag{3.9}$$

where q' is the N_e-dimensional vector given by

$$q' = q'(u^o, y) = (\mathbf{0}^T, R^{u^o,y}_{N_b} q^{N_b}, R^{u^o,y}_{N_e} q^{N_e})^T.$$

Substituting (3.9) into (3.8) yields

$$\overline{V}_t(q, x^i) = \min_{u^o \in \mathcal{N}_s} \min_{u^i \in \mathcal{I}^{x^i}} \sum_{y \in \mathcal{Y}} \min_{z \in \mathcal{Z}^{u^i}_{t+1}} (\bar{b}^{u^i,z}_{t+1})^T \left[\left(\widetilde{I\!P}^{u^i} \right)^T \mathcal{H}(D(\mathbf{R}^{u^o,y}) q) + q' \right],$$

which upon noting that $[\bar{b}^{u^i,z}_{t+1}]_{N_b} = c_b$ and $[\bar{b}^{u^i,z}_{t+1}]_{N_e} = c_e$,

$$= \min_{u^o \in \mathcal{N}_s} \min_{u^i \in \mathcal{I}^{x^i}} \sum_{y \in \mathcal{Y}} \min_{z \in \mathcal{Z}^{u^i}_{t+1}} \left[(\bar{b}^{u^i,z}_{t+1})^T \left(\widetilde{I\!P}^{u^i} \right)^T \mathcal{H}(D(\mathbf{R}^{u^o,y}) q) \right.$$

$$\left. + c_b R^{u^o,y}_{N_b} q^{N_b} + c_e R^{u^o,y}_{N_e} q^{N_e} \right]$$

$$= \min_{u^o \in \mathcal{N}_s} \min_{u^i \in \mathcal{I}^{x^i}} \sum_{y \in \mathcal{Y}} \left\{ \min_{z \in \mathcal{Z}^{u^i}_{t+1}} \left[(\bar{b}^{u^i,z}_{t+1})^T \left(\widetilde{I\!P}^{u^i} \right)^T \mathcal{H}(D(\mathbf{R}^{u^o,y}) q) \right] \right.$$

$$\left. + c_b R^{u^o,y}_{N_b} q^{N_b} + c_e R^{u^o,y}_{N_e} q^{N_e} \right\},$$

and noting that $\sum_{y \in \mathcal{Y}} R^{u^o,y}_j = 1$ for all $j \in \mathcal{N}_e$, this is

$$= \left\{ \min_{u^o \in \mathcal{N}_s} \min_{u^i \in \mathcal{I}^{x^i}} \sum_{y \in \mathcal{Y}} \min_{z \in \mathcal{Z}^{u^i}_{t+1}} \left[(\bar{b}^{u^i,z}_{t+1})^T \left(\widetilde{I\!P}^{u^i} \right)^T \mathcal{H}(D(\mathbf{R}^{u^o,y})q) \right] \right\}$$
$$+ c_b q^{N_b} + c_e q^{N_e}$$
$$= \left\{ \min_{u^o \in \mathcal{N}_s} \min_{u^i \in \mathcal{I}^{x^i}} \sum_{y \in \mathcal{Y}} \min_{z \in \mathcal{Z}^{u^i}_{t+1}} \left[(\widetilde{I\!P}^{u^i} \bar{b}^{u^i,z}_{t+1})^T \mathcal{H}(D(\mathbf{R}^{u^o,y})q) \right] \right\}$$
$$+ c_b q^{N_b} + c_e q^{N_e},$$

and letting $\tilde{\mathbf{R}}^{u^o,y}$ denote the first N_s elements of $\mathbf{R}^{u^o,y}$,

$$= \left\{ \min_{u^o \in \mathcal{N}_s} \min_{u^i \in \mathcal{I}^{x^i}} \sum_{y \in \mathcal{Y}} \min_{z \in \mathcal{Z}^{u^i}_{t+1}} \left[(D(\tilde{\mathbf{R}}^{u^o,y}) \widetilde{I\!P}^{u^i} \bar{b}^{u^i,z}_{t+1})^T \mathcal{H}(q) \right] \right\}$$
$$+ c_b q^{N_b} + c_e q^{N_e}.$$

□

Theorem 1. *Suppose there exist index sets $\mathcal{Z}^x_{t+1}$ (of cardinality Z^x_{t+1}) for all $x \in \mathcal{N}_b$, and $b^{x,z}_{t+1} \in I\!R^{N_s}$ for all $x \in \mathcal{N}_b$ and all $z \in \mathcal{Z}^x_{t+1}$ such that*

$$\overline{V}_{t+1}(q, x^i) = \min_{z \in \mathcal{Z}^{x^i}_{t+1}} \mathcal{G}(b^{x^i,z}_{t+1}) \cdot q \quad \forall q \in S^{N_e} \ \forall x^i \in \mathcal{N}_b.$$

Let $Z^{x^i}_t \doteq \sum_{u^i \in \mathcal{I}^{x^i}} |\mathcal{Z}^{u^i}_{t+1}|^{N_y} N^{x^i}_u$ and $\mathcal{Z}^{x^i}_t \doteq]1, Z^{x^i}_t[$. Let $\mathcal{M}$ be a one-to-one, onto mapping from $\mathcal{N}_s \times \mathcal{I}^{x^i} \times \bigcup_{u^i \in \mathcal{I}^{x^i}} \mathcal{P}^{N_y}\left(\mathcal{Z}^{u^i}_{t+1}\right) \to \mathcal{Z}^{x^i}_t$, where the notation $\mathcal{P}^N(\mathcal{Z})$ denotes the set of sequences of length N of elements of $\mathcal{Z}$. Lastly, let $\widetilde{I\!P}^{u^i}$ and $D(\tilde{\mathbf{R}}^{u^o,y})$ be as defined in Lemma 1. Then,

$$\overline{V}_t(q, x^i) = \min_{z \in \mathcal{Z}^{x^i}_t} \left[b^{x^i,z}_t \cdot \mathcal{H}(q) \right] + c_b q^{N_b} + c_e q^{N_e} \tag{3.10}$$
$$= \min_{z \in \mathcal{Z}^{x^i}_t} \mathcal{G}(b^{x^i,z}_t) \cdot q, \quad \forall q \in S^{N_e} \ \forall x^i \in \mathcal{N}_b, \tag{3.11}$$

where, for each $z \in \mathcal{Z}^{x^i}_t$,

$$b^{x^i,z}_t = \sum_{y \in \mathcal{Y}} D(\tilde{\mathbf{R}}^{u^o,y}) \widetilde{I\!P}^{u^i} \mathcal{G}(b^{u^i,z_y}_{t+1}), \tag{3.12}$$

and $(u^o, u^i, \{z_y\}_{y \in \mathcal{Y}}) = \mathcal{M}^{-1}(z)$.

Remark 1. It is important to note that this implies that the value function may be propagated purely by the algebraic operations given in (3.12). The set of vectors $\{b^{x,z}_t \,|\, z \in \mathcal{Z}^x_t,\, x \in \mathcal{N}_b\}$ completely define the function $\overline{V}_t$.

Proof. From Lemma 1, we have

$$\overline{V}_t(q,x^i)=\left\{\min_{u^o\in\mathcal{N}_s}\min_{u^i\in\mathcal{I}^{x^i}}\sum_{y\in\mathcal{Y}}\min_{z\in\mathcal{Z}^{u^i}_{t+1}}\left[(D(\tilde{\mathbf{R}}^{u^o,y})\widetilde{I\!P}^{u^i}\bar{b}^{u^i,z}_{t+1})^T\mathcal{H}(q)\right]\right\}$$
$$+c_bq^{N_b}+c_eq^{N_e}, \tag{3.13}$$

where we recall $\bar{b}^{x^i,z}_{t+1}\doteq\mathcal{G}(b^{x^i,z}_{t+1})$. It is handiest to employ the min-plus algebra form here. We remind the reader that the min-plus algebra (semifield) is defined on $I\!R\cup\{+\infty\}$, with addition and multiplication operations given by $a\oplus b\doteq\min\{a,b\}$ and $a\otimes b\doteq a+b$ (c.f., [1, 2, 11]). Using this formulation, (3.13) becomes

$$\overline{V}_t(q,x^i)=\left\{\bigoplus_{u^o\in\mathcal{N}_s}\bigoplus_{u^i\in\mathcal{I}^{x^i}}\bigotimes_{y\in\mathcal{Y}}\bigoplus_{z\in\mathcal{Z}^{u^i}_{t+1}}\left[\left(D(\tilde{\mathbf{R}}^{u^o,y})\widetilde{I\!P}^{u^i}\bar{b}^{u^i,z}_{t+1}\right)^T\mathcal{H}(q)\right]\right\}$$
$$+c_bq^{N_b}+c_eq^{N_e}.$$

Using the min-plus distributive property (c.f., [4, 5]), this becomes

$$\overline{V}_t(q,x^i)=\left\{\bigoplus_{u^o\in\mathcal{N}_s}\bigoplus_{u^i\in\mathcal{I}^{x^i}}\bigoplus_{\{z.\}\in\mathcal{P}^{N_y}(\mathcal{Z}^{u^i}_{t+1})}\bigotimes_{y\in\mathcal{Y}}\left[\left(D(\tilde{\mathbf{R}}^{u^o,y})\widetilde{I\!P}^{u^i}\bar{b}^{u^i,z_y}_{t+1}\right)^T\mathcal{H}(q)\right]\right\}$$
$$+c_bq^{N_b}+c_eq^{N_e}$$
$$=\left\{\bigoplus_{u^o\in\mathcal{N}_s}\bigoplus_{u^i\in\mathcal{I}^{x^i}}\bigoplus_{\{z.\}\in\mathcal{P}^{N_y}(\mathcal{Z}^{u^i}_{t+1})}\left[\sum_{y\in\mathcal{Y}}D(\tilde{\mathbf{R}}^{u^o,y})\widetilde{I\!P}^{u^i}\bar{b}^{u^i,z_y}_{t+1}\right]^T\mathcal{H}(q)\right\}$$
$$+c_bq^{N_b}+c_eq^{N_e}$$
$$=\left\{\min_{u^o\in\mathcal{N}_s}\min_{u^i\in\mathcal{I}^{x^i}}\min_{\{z.\}\in\mathcal{P}^{N_y}(\mathcal{Z}^{u^i}_{t+1})}\left[\sum_{y\in\mathcal{Y}}D(\tilde{\mathbf{R}}^{u^o,y})\widetilde{I\!P}^{u^i}\bar{b}^{u^i,z_y}_{t+1}\right]^T\mathcal{H}(q)\right\}$$
$$+c_bq^{N_b}+c_eq^{N_e}$$
$$=\left\{\min_{(u^o,u^i,\{z.\})\in\mathcal{N}_s\times\mathcal{I}^{x^i}\times\mathcal{P}^{N_y}(\mathcal{Z}^{u^i}_{t+1})}\left[\sum_{y\in\mathcal{Y}}D(\tilde{\mathbf{R}}^{u^o,y})\widetilde{I\!P}^{u^i}\bar{b}^{u^i,z_y}_{t+1}\right]^T\mathcal{H}(q)\right\}$$
$$+c_bq^{N_b}+c_eq^{N_e},$$

and using (3.12), this is

$$=\min_{z\in\mathcal{Z}^{x^i}_t}\left[b^{x^i,z}_t\cdot\mathcal{H}(q)\right]++c_bq^{N_b}+c_eq^{N_e}=\min_{z\in\mathcal{Z}^{x^i}_t}\mathcal{G}(b^{x^i,z}_t)\cdot q,$$

where $\mathcal{Z}^{x^i}_t$ is as asserted. □

It is worth noting that $\overline{V}_t$ is a concave, piecewise linear function of its first argument, the controller's probability distribution regarding the intruder position. For a problem with 8 UGSs in the region, a slice of $\overline{V}_t$ is depicted in Fig. 3.2, where $c_e=-100$ and $c_b=100$.

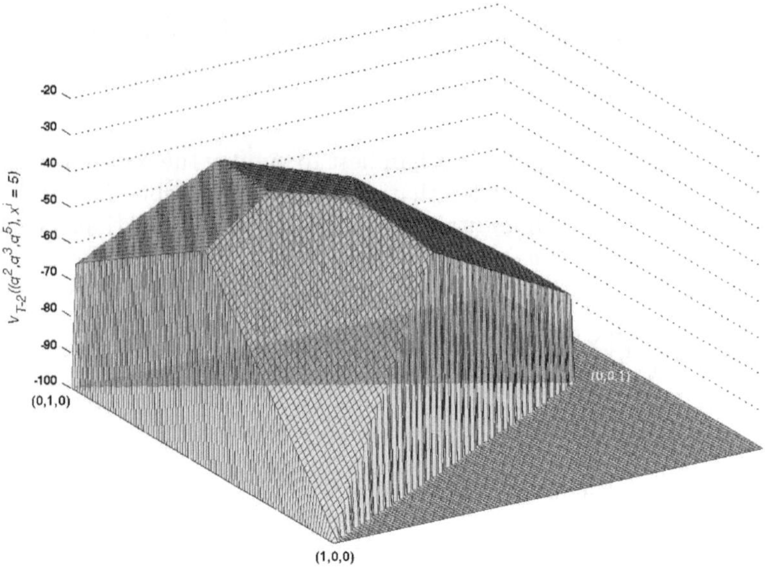

Fig. 3.2 An example of $\overline{V}_{T-2}(q, x^i)$ over a simplex slice in coordinates (q^2, q^3, q^5) with $x^i = 5$

3.4 Efficient Projection and Pruning

Although the curse-of-dimensionality is avoided with the idempotent-based dynamic program, another computational complexity issue arises, which may be referred to as the curse-of-complexity. Note that a naive gridding of the state-space imposes the curse-of-dimensionality where this might not actually be present in the solution. As trivial examples, note that a linear functional over $I\!R^N$ can be represented exactly with only N numbers, while using a grid-based representation (over only a rectangular region) would require a set of L^N numbers where L would be the number of grid-points per dimension. At the other extreme, a one-dimensional Brownian path would require a very large number of data points for reasonable approximate representation, even though $N = 1$. We see that complexity and dimension are not necessarily related, *and the curse-of-dimensionality is only an artifact of the classical representation form.* With certain idempotent-based algorithms, a low-complexity solution approximation might be obtained regardless of state-space dimension if the solution is indeed low-complexity. We do not make this more rigorous here.

The curse-of-complexity associated with the idempotent-based approach is evident in the extremely rapid growth of Z_t^x as one propagates backwards. In practice, it is often found that most of the linear functionals provide either zero additional value or very little. Also, [14], suggests that one should expect at worst polynomial growth rather than exponential.

In order to reduce the solution complexity, one should project the solution approximation, at each step, down onto a lower-dimensional min-plus subspace. It is known that the optimal projection is actually obtained through pruning of the set of linear functionals (c.f., [6]). Consequently, our optimal complexity attenuation problem reduces to an optimal pruning problem. In particular, one seeks the optimal projection onto a subspace of dimension, say N. Given the above, we see that this is obtained by optimally pruning the $\mathcal{Z}_t^x$ to a subset of size N. As the computations for selection of the optimal subset are costly, we apply a greedy (suboptimal) approach. One begins with the functional which obtains the minimum value over S^{N_e}. Given a subset of size $k < N$, one chooses the next functional from the remaining set, according to which additional functional would maximally reduce the pointwise minimum in an L_∞ sense. This process is repeated until one has a subset of size N.

There are two significant issues in this process. First, one must be able to evaluate the L_∞ change in the pointwise minimum which would be obtained by the inclusion of a new linear functional. This is more easily computed if one has the vertices of the piecewise linear, concave functional formed by the previously selected set of functionals. The second issue is the sequential computation of these vertices. We discuss each of these issues separately.

We begin with the first issue: the maximally-greedy, selection of the next linear functional to add to the current set. Here, the optimality is taken in terms of the L_∞ difference between the pointwise minimum of the set of k linear functionals and the pointwise minimum of the set of $k+1$ linear functionals obtained by the addition of one more functional. Motivation for selection of the L_∞ norm can be found in [6].

We now describe the greedy algorithm, and the computational efficiencies that may be obtained in implementation. Suppose we have $\overline{V}_t$ in the form (3.10), Let $e(q) \doteq c_b q^{N_b} + c_e q^{N_e}$, and note that we may then write $\overline{V}_t$ as

$$\overline{V}_t(q, x^i) = \min_{z \in \mathcal{Z}_t^{x^i}} \left[b_t^{x^i,z} \cdot \mathcal{H}(q) \right] + e(q) \doteq \widetilde{V}_t(\mathcal{H}(q), x^i) + e(q). \tag{3.14}$$

Fix any x^i. Suppose we wish to find an approximately optimal projection/pruning of $\widetilde{V}_t(\cdot, x^i)$ with complexity $N < Z_t^{x^i}$, that is an approximation

$$\widetilde{V}_t^N(\mathcal{H}(q), x^i) = \min_{z \in \tilde{\mathcal{Z}}} \left[b_t^{x^i,z} \cdot \mathcal{H}(q) \right] \quad \forall\, q \in S^{N_e}, \tag{3.15}$$

where $|\tilde{\mathcal{Z}}| = N$. Again, by [6] the optimal projection is obtained by pruning. The maximally-greedy (suboptimal) algorithm we employ is as follows. First, we select

$$z_1 \in \operatorname{argmin} \left\{ \min_{q \in S^{N_s}} \left[b_t^{x^i,z_1} \cdot q \right] \,\middle|\, z \in \mathcal{Z}_t^{x^i} \right\},$$

and let $\mathcal{Z}'_1 = \{z_1\}$. Given $\mathcal{Z}'_k$, we select z_{k+1}, and let $\mathcal{Z}'_{k+1} = \mathcal{Z}'_k \cup \{z'_{k+1}\}$, as follows. Let

$$\begin{aligned}
z_{k+1} &\in \operatorname{argmax}\Big\{ \max_{q\in S^{N_e}} \Big[\min_{z\in\mathcal{Z}'_k}(b_t^{x^i,z}\cdot\mathcal{H}(q)) - \min_{z\in\mathcal{Z}'_k\cup\{z\}}(b_t^{x^i,z}\cdot\mathcal{H}(q))\Big] \,\Big|\, z\in\mathcal{Z}_t^{x^i}\Big\} \\
&= \operatorname{argmax}\Big\{ \max_{q\in S^{N_s}} \Big[\min_{z\in\mathcal{Z}'_k}(b_t^{x^i,z}\cdot q) - \min_{z\in\mathcal{Z}'_k\cup\{z\}}(b_t^{x^i,z}\cdot q)\Big] \,\Big|\, z\in\mathcal{Z}_t^{x^i}\Big\} \\
&= \operatorname{argmin}\Big\{ \min_{q\in S^{N_s}} \Big[\min_{z\in\mathcal{Z}'_k\cup\{z\}}(b_t^{x^i,z}\cdot q) - \min_{z\in\mathcal{Z}'_k}(b_t^{x^i,z}\cdot q)\Big] \,\Big|\, z\in\mathcal{Z}_t^{x^i}\Big\} \\
&\doteq \mathcal{Z}'_{k+1}.
\end{aligned}$$

Suppose

$$\min_{q\in S^{N_e}} \Big[\min_{z\in\mathcal{Z}'_k\cup\{z\}}(b_t^{x^i,z}\cdot q) - \min_{z\in\mathcal{Z}'_k}(b_t^{x^i,z}\cdot q)\Big] < 0, \tag{3.16}$$

i.e., that there is a $z\in\mathcal{Z}_t^{x^i}\setminus\mathcal{Z}'_k$ which improves the pointwise minimum. (Otherwise, nothing will be gained by the addition of more functionals.) Then

$$\begin{aligned}
\mathcal{Z}'_{k+1} &= \operatorname{argmin}\Big\{ \min_{q\in S^{N_s}} \Big[b_t^{x^i,z}\cdot q - \min_{z\in\mathcal{Z}'_k}(b_t^{x^i,z}\cdot q)\Big] \,\Big|\, z\in\mathcal{Z}_t^{x^i}\Big\} \\
&\doteq \operatorname{argmin}\Big\{ \min_{q\in S^{N_s}} \Big[b_t^{x^i,z}\cdot q - \gamma(q,\mathcal{Z}'_k)\Big] \,\Big|\, z\in\mathcal{Z}_t^{x^i}\Big\}.
\end{aligned} \tag{3.17}$$

Now we examine algorithms for efficient computation of the right-hand side of (3.17). For $\mathcal{Z}'_k\subseteq\mathcal{Z}_t^{x^i}$, let $\mathcal{V}_k=\mathcal{V}_k(\mathcal{Z}'_k)$ be the set of vertices generated by hyperplanes associated to the linear functionals $b_t^{x^i,z}\cdot q$ for $z\in\mathcal{Z}'_k$ as well as the boundary planes of S^{N_s}. Let $\mathcal{Q}(\mathcal{V}_k)$ be the set of $q\in S^{N_s}$ associated to each vertex in $\mathcal{V}_k$, i.e.,

$$\mathcal{Q}(\mathcal{V}_k) \doteq \left\{ q\in S^{N_s} \,|\, (q,\gamma(q,\mathcal{Z}'_k))\in\mathcal{V}_k \right\}.$$

It is not difficult to see that

$$\mathcal{Z}'_{k+1} = \operatorname{argmin}\Big\{ \min_{q\in\mathcal{Q}(\mathcal{V}_k)} \Big[b_t^{x^i,z}\cdot q - \gamma(q,\mathcal{Z}'_k)\Big] \,\Big|\, z\in\mathcal{Z}_t^{x^i}\Big\}. \tag{3.18}$$

Note that

$$\begin{aligned}
&\min_{z\in\mathcal{Z}_t^{x^i}} \min_{q\in\mathcal{Q}(\mathcal{V}_k)} \Big[b_t^{x^i,z}\cdot q - \gamma(q,\mathcal{Z}'_k)\Big] \\
&= \min_{q\in\mathcal{Q}(\mathcal{V}_k)} \Big\{ \min_{z\in\mathcal{Z}_t^{x^i}} \big[b_t^{x^i,z}\cdot q\big] - \gamma(q,\mathcal{Z}'_k)\Big\},
\end{aligned}$$

which by Theorem 1,

$$\begin{aligned}
&= \min_{q\in\mathcal{Q}(\mathcal{V}_k)} \Big\{ \min_{(u^o,u^i)\in\mathcal{N}_s\times\mathcal{I}^{x^i}} \min_{\{\zeta.\}\in\mathcal{P}^{N_y}(\mathcal{Z}_{t+1}^{u^i})} \Big[\sum_{y\in\mathcal{Y}} \tilde{b}_t^{x^i,u^o,u^i,\zeta_y,y}\cdot q\Big] \\
&\qquad -\gamma(q,\mathcal{Z}'_k)\Big\},
\end{aligned} \tag{3.19}$$

where $\tilde{b}_t^{x^i,u^o,u^i,\zeta_y,y} = \sum_{y\in\mathcal{Y}} D(\tilde{\mathbf{R}}^{u^o,y})\widetilde{I\!P}^{u^i}\mathcal{G}(b_{t+1}^{u^i,\zeta_y})$ for all $x^i \in \mathcal{N}_e$, all $(u^o,u^i) \in \mathcal{N}_s \times \mathcal{I}^{x^i}$ and all $(\{\zeta_\cdot\},y) \in \mathcal{P}^{N_y}(\mathcal{Z}_{t+1}^{u^i}) \times \mathcal{Y}$.

Theorem 2. *Given finite index set $\mathcal{Z}$, $x^i \in \mathcal{N}_b$, $(u^o,u^i) \in \mathcal{N}_s \times \mathcal{I}^{x^i}$, $q \in S^{N_s}$ and $y \in \mathcal{Y}$, let*

$$\zeta_y = \zeta_y^{u^o,u^i}(q,\mathcal{Z}) \in \operatorname*{argmin}_{\zeta\in\mathcal{Z}} \tilde{b}_t^{x^i,u^o,u^i,\zeta,y} \cdot q. \tag{3.20}$$

Also let

$$\widehat{\mathcal{Z}}^{u^o,u^i}(q,\mathcal{Z}) \doteq \left\{\{\zeta_\cdot\} = \{\zeta_y\}_{y\in\mathcal{Y}} \in \mathcal{P}^{N_y}(\mathcal{Z}) \,\middle|\, \zeta_y \textit{ satisfies (3.20) } \forall y \in \mathcal{Y}\right\}. \tag{3.21}$$

Then, for any finite index set $\mathcal{Z}$, $x^i \in \mathcal{N}_b$, $(u^o,u^i) \in \mathcal{N}_s \times \mathcal{I}^{x^i}$ and $q \in S^{N_s}$,

$$\min_{\{\zeta_\cdot\}\in\mathcal{P}^{N_y}(\mathcal{Z})} \left[\sum_{y\in\mathcal{Y}} \tilde{b}_t^{x^i,u^o,u^i,\zeta_y,y} \cdot q\right] = \min_{\{\zeta_\cdot\}\in\widehat{\mathcal{Z}}^{u^o,u^i}(q,\mathcal{Z})} \left[\sum_{y\in\mathcal{Y}} \tilde{b}_t^{x^i,u^o,u^i,\zeta_y,y} \cdot q\right].$$

Proof. Fix index set $\mathcal{Z}$, $(u^o,u^i) \in \mathcal{N}_s \times \mathcal{I}^{x^i}$ and $q \in S^{N_s}$. By set inclusion, it is immediate that

$$\min_{\{\zeta_\cdot\}\in\mathcal{P}^{N_y}(\mathcal{Z})} \left[\sum_{y\in\mathcal{Y}} \tilde{b}_t^{x^i,u^o,u^i,\zeta_y,y} \cdot q\right] \le \min_{\{\zeta_\cdot\}\in\widehat{\mathcal{Z}}^{u^o,u^i}(q,\mathcal{Z})} \left[\sum_{y\in\mathcal{Y}} \tilde{b}_t^{x^i,u^o,u^i,\zeta_y,y} \cdot q\right].$$

We prove the reverse inequality. Let $\{\bar{\zeta}_\cdot\} = \{\bar{\zeta}_y\}_{y\in\mathcal{Y}} \in \mathcal{P}^{N_y}(\mathcal{Z})$ and $\{\hat{\zeta}_\cdot\} = \{\hat{\zeta}_y\}_{y\in\mathcal{Y}} \in \mathcal{Z}^{u^o,u^i}(q,\mathcal{Z})$ By (3.21), for each $y \in \mathcal{Y}$

$$\tilde{b}_t^{x^i,u^o,u^i,\bar{\zeta}_y,y} \cdot q \ge \tilde{b}_t^{x^i,u^o,u^i,\hat{\zeta}_y,y} \cdot q,$$

which implies

$$\sum_{y\in\mathcal{Y}} \tilde{b}_t^{x^i,u^o,u^i,\bar{\zeta}_y,y} \cdot q. \ge \sum_{y\in\mathcal{Y}} \tilde{b}_t^{x^i,u^o,u^i,\hat{\zeta}_y,y} \cdot q$$

Since this is true for all $\{\bar{\zeta}_\cdot\} \in \mathcal{P}^{N_y}(\mathcal{Z})$,

$$\begin{aligned}\min_{\{\zeta_\cdot\}\in\mathcal{P}^{N_y}(\mathcal{Z})} \left[\sum_{y\in\mathcal{Y}} \tilde{b}_t^{x^i,u^o,u^i,\bar{\zeta}_y,y} \cdot q\right] &\ge \sum_{y\in\mathcal{Y}} \tilde{b}_t^{x^i,u^o,u^i,\hat{\zeta}_y,y} \cdot q \\ &\ge \min_{\{\zeta_\cdot\}\in\widehat{\mathcal{Z}}^{u^o,u^i}(q,\mathcal{Z})} \left[\sum_{y\in\mathcal{Y}} \tilde{b}_t^{x^i,u^o,u^i,\hat{\zeta}_y,y} \cdot q\right]. \quad \square\end{aligned}$$

Employing Theorem 2 in (3.19) yields

$$\min_{z\in\mathcal{Z}_t^{x^i}} \min_{q\in\mathcal{Q}(\mathcal{V}_k)} \left[b_t^{x^i,z}\cdot q-\gamma(q,\mathcal{Z}_k')\right]$$

$$=\min_{q\in\mathcal{Q}(\mathcal{V}_k)}\Big\{\min_{(u^o,u^i)\in\mathcal{N}_s\times\mathcal{I}^{x^i}} \min_{\{\zeta_\cdot\}\in\widehat{\mathcal{Z}}^{u^o,u^i}(q,\mathcal{Z}_{t+1}^{u^i})}\Big[\sum_{y\in\mathcal{Y}}\tilde{b}_t^{x^i,u^o,u^i,\zeta_y,y}\cdot q\Big]-\gamma(q,\mathcal{Z}_k')\Big\},$$

and letting $\hat{b}_t^{x^i,u^o,u^i,\{\zeta_\cdot\}}\doteq\sum_{y\in\mathcal{Y}}\tilde{b}_t^{x^i,u^o,u^i,\zeta_y,y}$, this is

$$=\min_{q\in\mathcal{Q}(\mathcal{V}_k)}\Big\{\min_{(u^o,u^i)\in\mathcal{N}_s\times\mathcal{I}^{x^i}} \min_{\{\zeta_\cdot\}\in\widehat{\mathcal{Z}}^{u^o,u^i}(q,\mathcal{Z}_{t+1}^{u^i})}\hat{b}_t^{x^i,u^o,u^i,\{\zeta_\cdot\}}\cdot q-\gamma(q,\mathcal{Z}_k')\Big\}. \quad (3.22)$$

Suppose the minimum on (3.22) is achieved at $q^*,u_*^o,u_*^i,\{\zeta_\cdot^*\}$. Then, one lets $z_{k+1}=\mathcal{M}^{-1}(u_*^o,u_*^i,\{\zeta_\cdot^*\})$ and $\mathcal{Z}_{k+1}'=\mathcal{Z}_k'\cup\{z_{k+1}\}$. We continue the procedure until $k+1=N$ (or until the left-hand side of (3.16) is zero). The resulting set of indices is $\tilde{\mathcal{Z}}$. We let $\widetilde{V}_t^N$ be given by (3.15). As in (3.14),(3.15), for each $x^i\in\mathcal{N}_e$ we obtain the approximation

$$\begin{aligned}\overline{V}_t(q,x^i)\simeq\overline{V}_t^N(q,x^i)&\doteq\widetilde{V}_t^N(\mathcal{H}(q),x^i)+e(q)\\ &=\min_{z\in\tilde{\mathcal{Z}}}\left[b_t^{x^i,z}\cdot\mathcal{H}(q)\right]+c_bq^{N_b}+c_eq^{N_e}. \quad (3.23)\end{aligned}$$

For further backward propagation, we could replace $\mathcal{Z}_t^{x^i}$ with $\tilde{\mathcal{Z}}$, and $\overline{V}_t$ with $\overline{V}_t^N(q,x^i)$. Each step in the resulting approximate, idempotent form of the dynamic programming consists of two parts: the backward propagation according to Theorem 1, followed by the projection down to a set of cardinality N. Let the backward propagation via idempotent representation operator (Theorem 1) be denoted by $\mathcal{S}$. That is, $\overline{V}_t=\mathcal{S}[\overline{V}_{t+1}]$. Let the above min-plus projection operator onto N-dimensional min-plus subspace be denoted by Π^N, where above, this is $\overline{V}_t^N=\Pi^N[\overline{V}_t]$. Together, the approximate idempotent backward dynamic program is

$$\widehat{V}_t=\Pi^N\big[\mathcal{S}[\widehat{V}_{t+1}]\big], \quad (3.24)$$

$$\widehat{V}_T(q,x^i)=\bar{\beta}\cdot q\quad\forall q\in S^{N_e},\ \forall x^i\in\mathcal{N}_e. \quad (3.25)$$

The efficient implementation of the projection substep of the above algorithm requires an ability to sequentially produce the vertices of the convex polytope generated by the $\mathcal{Z}_k'$ as one iterates over $k\in]1,N[$. More specifically, the polytope is the hypograph of $\gamma(\cdot,\mathcal{Z}_k')$ over S^{N_s}. The sequential computation of this set of vertices is very efficiently performed by the double description method [12]. Let us describe the double description approach to convex polytope representation a bit. Note that the polytope may be uniquely represented either by the set of hyperplanes (more exactly, half-spaces) defining it, or by its set of vertices. The key to the algorithm is that one keeps track of both the set of half-spaces and the set of vertices at each step. The half-spaces defining the hypograph of $\gamma(\cdot, \mathcal{Z}'_k)$ are those indexed by the functionals, i.e., the $z \in \mathcal{Z}'_k$, plus the half-space boundaries of S^{N_s}. If we let $w \in I\!R$ denote the range dimension and $q' \in S^{N_s} \subset I\!R^{N_s}$ denote the domain dimensions, each $z_j \in \mathcal{Z}'_k$ corresponds to a half-space $w - b_t^{x^i, z_j} \cdot q' \leq 0$. (For our purposes, we do not need to include a "bottom" for the polytope - the interior is unbounded.) At each additional z_{k+1}, we are adding a half-space constraint, $w - b_t^{x^i, z_{k+1}} \cdot q' \leq 0$, with corresponding face $w - b_t^{x^i, z_{k+1}} \cdot q' = 0$. Recall the set of vertices, prior to the additional z_{k+1}, is denoted by $\mathcal{V}_k$. One needs to determine the resulting new set of vertices, $\mathcal{V}_{k+1}$, efficiently. The double description method does this.

The hyperplane $w - b_t^{x^i, z_{k+1}} \cdot q' = 0$ partitions $\mathcal{V}_k$ into two subsets, $\mathcal{V}_k^+$ and $\mathcal{V}_k^-$ where $\mathcal{V}_k^- = \{(q'_\alpha, w_\alpha) \in \mathcal{V}_k \,|\, w_\alpha - b_t^{x^i, z_{k+1}} \cdot q'_\alpha < 0\}$ and $\mathcal{V}_k^+ = \{(q'_\alpha, w_\alpha) \in \mathcal{V}_k \,|\, w_\alpha - b_t^{x^i, z_{k+1}} \cdot q'_\alpha > 0\}$. We ignore the special case of equality in our brief discussion here. Those vertices in $\mathcal{V}_k^+$ are dropped; they are outside the new polytope. The vertices in $\mathcal{V}_k^-$ are retained. Where there is $(q'_\alpha, w_\alpha) \in \mathcal{V}_k^-$ and $(q'_\beta, w_\beta) \in \mathcal{V}_k^+$ which are joined by an edge of the polytope, one must compute the resulting new vertex formed by the intersection of that edge and the new hyperplane $w - b_t^{x^i, z_{k+1}} \cdot q' = 0$. The union of this set of new vertices and $\mathcal{V}_k^-$ is $\mathcal{V}_{k+1}$.

3.4.1 $\mathcal{L}_\infty$ Error Bounds

Recall from Theorem 1 and (3.24), that the ideal idempotent backward propagation is

$$\overline{V}_t = \mathcal{S}[\overline{V}_{t+1}],$$

and the approximate idempotent backward propagation is

$$\widehat{V}_t = \Pi^N\big[\mathcal{S}[\widehat{V}_{t+1}]\big],$$

both with the same terminal condition, $\widehat{V}_T(q, x^i) = \overline{V}_T(q, x^i) = \bar{\beta} \cdot q$. Let the one-step projection error be

$$e_t^N \doteq \big\|\widehat{V}_t - \mathcal{S}[\widehat{V}_{t+1}]\big\|_\infty = \big\|\Pi^N\big[\mathcal{S}[\widehat{V}_{t+1}]\big] - \mathcal{S}[\widehat{V}_{t+1}]\big\|_\infty,$$

where $\|\widehat{V}_t\|_\infty \doteq \max_{x \in \mathcal{N}_e} \sup_{q \in S^{N_e}} |\widehat{V}_t(q, x)|$. Let the total error be denoted as $\bar{e}_t \doteq \|\widehat{V}_t - \overline{V}_t\|_\infty$. Then,

$$\begin{aligned}\bar{e}_t &= \|\widehat{V}_t - \overline{V}_t\|_\infty \\ &\leq \|\widehat{V}_t - \mathcal{S}[\widehat{V}_{t+1}]\|_\infty + \|\mathcal{S}[\widehat{V}_{t+1}] - \overline{V}_t\|_\infty \\ &= \|\widehat{V}_t - \mathcal{S}[\widehat{V}_{t+1}]\|_\infty + \|\mathcal{S}[\widehat{V}_{t+1}] - \mathcal{S}[\overline{V}_{t+1}]\|_\infty \\ &= e_t^N + \|\mathcal{S}[\widehat{V}_{t+1}] - \mathcal{S}[\overline{V}_{t+1}]\|_\infty\end{aligned}$$

which by Theorem 4.2 in [13],

$$\leq e_t^N + \|\widehat{V}_{t+1} - \overline{V}_{t+1}\|_\infty = e_t^N + \bar{e}_{t+1}.$$

Applying an induction argument, one finds

$$\bar{e}_t \leq \sum_{r=t}^{T-1} e_r^N.$$

This is, of course, a conservative bound on the error growth rate. If the worst-case projection errors occurred at different (q, x^i) locations at different steps, then the errors would not grow additively. Secondly, the probability mass moves toward nodes N_b and N_e (where there is no error in the value) as one propagates forward, and so we can expect reduction in error effects from this.

3.5 Examples

In order to give a sense of the value of the above tools as well as their computational tractability, we exercise the tools on an example. The results appear in Table 3.3. In particular, we apply the above techniques to the problem depicted in Figure 3.1, where there are 8 UGSs, one UAV, one interceptor and one intruder. The road network and UGS locations are exactly as depicted in the figure. The payoffs are -100 points if the interceptor is co-located with the intruder prior to the intruder reaching the base ($c_e = -100$), 100 points if the intruder gets to the base ($c_b = 100$), and zero otherwise. The interceptor initial position is indicated in column 1. The initial probability distribution regarding the intruder is uniform over the locations other than N_b, N_e and x^i, where we note that $x^i = N_b$ in the last row. The final column is the value function at time $T - 7$, computed by the above algorithm, with $N = 20$ functional hyperplanes retained at each step. The second column is a check on the value function computation, obtained by applying the optimal control obtained during the value function (idempotent backward dynamic program) computation. Comparing the second and last columns in Table 3.3, we see that there is good correspondence between the Monte Carlo results and the computed value. The number of runs in the Monte Carlo tests per data point is $300,000$. The initial intruder position was generated according to the uniform distribution indicated above. The probability transition matrix, $\mathbb{P}$, which was used in the testing is given in Table 3.1. Regarding observation errors, the false-negative and false-positive probabilities are given

Table 3.1 Probability transition matrix

x_t^R	Next intruder postion, x_{t+1}^R 1	2	3	4	5	6	7	8	Base
1	0	0.20	0.45	0.35	0	0	0	0	0
2	0.10	0.30	0	0	0.60	0	0	0	0
3	0	0	0	0.45	0.25	0.30	0	0	0
4	0.15	0	0.15	0.10	0	0.60	0	0	0
5	0	0.10	0.25	0	0.05	0	0.35	0.25	0
6	0	0	0.30	0.10	0	0.02	0.58	0	0
7	0	0	0	0	0.10	0.20	0	0.40	0.30
8	0	0	0	0	0.10	0	0.15	0	0.75
Base	0	0	0	0	0	0	0	0	1

Table 3.2 Confusion matrix

UAV control, u^o	Probability False negative	False positive
1	0.044	0.013
2	0.029	0.011
3	0.015	0.016
4	0.018	0.010
5	0.038	0.012
6	0.060	0.001
7	0.008	0.008
8	0.004	0.018

in Table 3.2. In particular, the second column of Table 3.2 is the probability of observing $y_t = 0$ at node u_t^o (indicated in the first column) given $x_t^R = u_t^o$. The third column is the probability of observing $y_t = 1$ at node u_t^o given $x_t^R \neq u_t^o$ (assumed independent of the actual location $x_t^R \neq u_t^o$).

Columns 3 and 4 of Table 3.3 indicate the expected payoffs for two heuristic controllers, also obtained by the Monte Carlo method with the same number of runs. The first heuristic controller is as follows. At each step, the UAV moves to the location with the highest a priori probability of containing the intruder. The interceptor moves to the most likely next position of the intruder based on the current observation-conditioned distribution and the probability transition matrix. The results with this heuristic appear in column 3, and are substantially inferior to those obtained with the computed optimal controls. Note from Figure 3.1, that the intruder must pass through either node 7 or node 8 just prior to reaching the base. The second heuristic controller exploits this. In the second heuristic controller, the interceptor oscillates between nodes 7 and 8, while the UAV acts exactly as in the first heuristic. In the case where the initial interceptor position is 6, we let the

Table 3.3 Monte Carlo testing results

Initial interceptor postion, x^i	Payoff			
	Monte Carlo	Heuristic 1	Heuristic 2	DP
1	-7.7	23.9	N/A	-6.9
2	-25.8	8.4	N/A	-26.1
3	-26.7	17.7	N/A	-26.4
4	-14.5	20.8	N/A	-14.2
5	-39.7	-4.9	1.2	-39.4
6	-33.6	5.8	8.5	-33.2
7	-50.4	4.1	4.3	-50.1
8	-54.0	-13.5	-9.5	-53.7
Base	-40.6	-12.5	-1.3	-40.0

interceptor move first to node 7, and then begin oscillating. In the case where the initial interceptor position is 5 or $9 = N_b$, we let the interceptor move first to node 8, and then begin oscillating. We did not apply the second heuristic in cases where the initial interceptor position was 1, 2, 3 or 4. Clearly, the second heuristic controller also substantially underperforms relative to the computed optimal controls.

3.6 Closing Comments

A high-dimensional partially-observed stochastic control problem with observation-process control has been defined and motivated. In addition to the UAV observation process, the interceptor also provides both physical-domain control and observational data, and so has a dual-control nature. The Separation Principle is applied to create a control problem formulation with state space which is essentially $S^{N_s} \times \mathcal{N}_s$. This is a high-dimensional problem for moderately-sized examples, and so classical methods may not be computationally tractable.

A min-plus curse-of-dimensionality-free algorithm is developed for the problem. The associated curse-of-complexity is attenuated through repeated projection to a low-dimensional min-plus subspace. The optimal projection is obtained by pruning, and a greedy pruning algorithm, computationally enhanced by use of the double description method, is developed.

This has been, so far, demonstrated to perform well on problems with up to ten UGSs ($N_s = 10$). Here, an example with $N_s = 8$ is presented. We see that the resulting controls greatly outperform a reasonable heuristic. It is shown that the errors grow at most linearly with time-horizon length. However, that bound is clearly excessively conservative for long time horizons.

The actual per-step errors expected as a function of representation complexity are unknown, and likely problem-dependent. Development of an approach to estimation of these errors is one open problem. The possibility of additional computational efficiencies is another open problem. On a more theoretical note, demonstration of the applicability of the Separation Principle (or counterexample to such) for this problem class would be a useful advance.

References

1. Baccelli, F.L., Cohen, G., Olsder, G.J., Quadrat, J.-P.: Synchronization and Linearity. John Wiley, New York (1992)
2. Cuninghame-Green, R.A.: Minimax Algebra. Lecture Notes in Economics and Mathematical Systems, vol. 166. Springer, New York (1979)
3. Kalyanam, K., Dharba, S., Pachter, M., Chandler, P.: Optimal Search for a Moving Ground Target (2011) (preprint)
4. McEneaney, W.M., Kaise, H., Han, S.H.: Idempotent Method for Continuous-Time Stochastic Control and Complexity Attenuation. In: Proc. 2011 IFAC (2011)
5. McEneaney, W.M.: Idempotent algorithms for discrete-time stochastic control. Automatica 47, 443–451 (2011)
6. McEneaney, W.M.: "Complexity reduction, cornices and pruning. In: Litvinov, G.L., Sergeev, S.N. (eds.) Proc. of the International Conference on Tropical and Idempotent Mathematics. Contemporary Math., vol. 495, pp. 293–303. Amer. Math. Soc. (2009)
7. McEneaney, W.M., Deshpande, A., Gaubert, S.: Curse-of-complexity attenuation in the curse-of-dimensionality-free method for HJB PDEs. In: Proc. Amer. Control Conf. (2008)
8. McEneaney, W.M.: A curse-of-dimensionality-free numerical method for solution of certain HJB PDEs. SIAM J. on Control and Optim. 46, 1239–1276 (2007)
9. McEneaney, W.M., Oran, A., Cavender, A.: Value-Based Tasking Controllers for Sensing Assets. In: AIAA Guidance, Navigation and Control Conf., Honolulu (2008)
10. McEneaney, W.M., Singh, R.: Robustness Against Deception. In: Kott, A., McEneaney, W.M. (eds.) Adversarial Reasoning: Computational Approaches to Reading the Opponent's Mind, pp. 167–208. Chapman and Hall/CRC Press (2007)
11. McEneaney, W.M.: Max-Plus Methods for Nonlinear Control and Estimation. Birkhauser, Boston (2006)
12. Motzkin, T.S., Raiffa, H., Thompson, G.L., Thrall, R.M.: The double description method. Contributions to the Theory of Games II, 51–73 (1953)
13. Oran, A., McEneaney, W.M.: Max-Plus Enabled Dynamic Programming for Sensor Plaform Tasking (submitted)
14. Sontag, E.D.: VC dimension of neural networks. In: Bishop, C.M. (ed.) Neural Networks and Machine Learning, pp. 69–95. Springer (1998)

Chapter 4
Model Checking for Verification in UAV Cooperative Control Applications

Laura R. Humphrey

Abstract. Recently, there has been an increased interest in verification techniques for complex, autonomous systems. Consider for instance a scenario in which individual autonomous agents must work together to complete a series of constrained tasks. In cases such as this, there is a need for techniques to verify that the agents interact correctly, i.e. that they are able to cooperatively complete the required tasks while satisfying the constraints. Current research suggests that standard software modeling and verification techniques such as model checking can be extended to meet this need. Here, we explore the use of model checking for verification in three scenarios that include a team of unmanned aerial vehicles. The first scenario involves a centralized cooperative control scheme, the second involves a decentralized cooperative control scheme, and the third involves high-level mission planning. For these scenarios, tasks and constraints are written in linear temporal logic, scenario models are coded in PROMELA, and the models are verified using the model checker SPIN.

4.1 Introduction

Autonomous multi-agent systems have been the focus of a great deal of research spanning the course of many years [21, 45]. Such systems are in high demand due to their potential to solve complex problems quickly, to enable new classes of system capabilities, to increase manpower efficiencies, and to reduce overall costs. Indeed, the United States Air Force has expressed a strong desire to continue the development of autonomous multi-agent systems in its 2010 *Report on Technology Horizons* [9]. However, the same report notes a barrier to the practical use of autonomous systems:

> It is possible to develop systems having high levels of autonomy, but it is the lack of V&V methods that prevents all but relatively low levels of autonomy from being certified for use.

Laura R. Humphrey
AFRL Control Science Center of Excellence, WPAFB OH
e-mail: `laura.humphrey@wpafb.af.mil`

F. Fahroo et al. (Eds.): *Recent Adv. in Res. on Unmanned Aerial Veh.*, LNCIS 444, pp. 69–117.
DOI: 10.1007/978-3-642-37694-8_4

Because autonomous systems are primarily implemented as software, a solution to this problem could lie in software modeling and design techniques. As the level of complexity of software has increased, so has the sophistication of these techniques, as described in Ref. [5]. To summarize, the advent of computers saw hardware that was much more expensive than software. Software was short, highly specialized, and carefully coded. As hardware costs decreased relative to software costs and modifications to software became easier, a "code and fix" approach emerged that resulted in heavily patched and fragile software. Approaches such as requirements engineering, formal methods [16], and structured programming were promoted in order to address this issue. Of particular note is Royce's "waterfall" model [33], an iterative and incremental method that divides the system development process into increasingly more detailed stages: system requirements, software requirements, preliminary design, analysis, program design, coding, testing, and operation.

As software grew in size and complexity, new tools and processes were created for organizing large-scale efforts. In addition, the emergence of object-oriented programming allowed for more conceptually organized and modularized code, enabling better methods for modeling and documentation. Recent years have seen a continued focus on development practices, examining issues such as how to organize large projects, how to make use of information technology, how to reduce development costs, and how to increase system quality and dependability. The ever-increasing complexity of software projects as well as rapidly changing requirements has also resulted in a push toward model-driven development, in which standard models of a domain of interest are developed, allowing systems engineering to be used in conjunction with software engineering.

In summary, trends in software development include better organization and documentation, more requirements analysis, and an increased emphasis on high-level and model-driven design. These trends have been useful in containing costs and improving quality as the size and complexity of software has grown. Furthermore, studies of coding projects have repeatedly shown that a significant portion of programming errors can be traced back to errors in system requirements and high-level designs [3, 4, 40], and errors that are discovered earlier in the design process are cheaper to fix [40, 41]. Thus, it appears that in developing and promoting verification techniques for autonomous systems, there is value in concentrating on the early requirements analysis and preliminary design phases.

Here, we explore the use of formal methods for design and verification of cooperative multi-agent systems, specifically those that include unmanned aerial vehicles (UAVs). According to [2], a report written by the FAA and NASA concludes that

> Formal methods should be part of the education of every computer scientist and software engineer, just as the appropriate branch of applied maths is a necessary part of the education of all other engineers.

Formal methods can be divided into two categories: model checking and theorem proving [12]. Here, the focus will be on model checking. In model checking, a finite-state model of a system design is developed, a set of desired system properties is specified, and a piece of automated software called a *model checker* performs an exhaustive analysis of the model to verify that it has all the desired properties.

Desired properties can be expressed in several ways, for instance by using linear temporal logic (LTL) [31]. LTL is a modal logic in which formulas contain atomic propositions related by Boolean and temporal operators. It assumes a linear view of time; i.e., the system state at any discrete point in time will have a single successor state, even if that state is one several possible outcomes. Another modal logic commonly used for expressing desired properties is computation tree logic (CTL) [2]. CTL takes a branching view of time; i.e., the system state at any discrete point in time can have many possible successors, all of which are reasoned about collectively. Thus, in addition to the operators used in LTL, CTL includes two existential quantifiers, $\exists$ "exists" and $\forall$ "for all." The first indicates that a property should hold for at least one possible series of successor states or computation paths in the model, and the second indicates that a property should hold for all possible paths. Each logic has its own advantages and disadvantages [43].

LTL, CTL, and model checking have been used recently in the development of simple multi-UAV systems [24, 39]. Here, we explore the use of LTL, the SPIN model checker [17], and the modeling language PROMELA [13] for high-level design and verification in three UAV related applications. The first involves a centralized UAV cooperator controller that coordinates the actions of multiple UAVs performing a monitoring task. The second involves a leader election protocol for a decentralized system of unattended ground sensors sending estimates of an intruder's position to a UAV. The third involves verification of high-level UAV mission plans for a scenario in which multiple UAVs must be used to safely escort a "very important person" (VIP) across a road network. Sect. 4.2 presents the necessary background on LTL, the design language PROMELA, and the model checker SPIN; Sect. 4.3 describes the design and verification process in each of the three applications; Sect. 4.4 discusses the advantages and disadvantages of using model checking for such applications; and Sect. 4.5 gives an overview of current and future research directions.

4.2 Background

PROMELA[19] (Process or Protocol Meta Language) is a modeling language designed primarily for the description of asynchronous processes, especially those involved in software and hardware design. For ease in modeling realistic systems, it includes constructs for explicitly modeling non-deterministic behavior. The model checker SPIN (Simple PROMELA Interpreter) is "a generic verification system that supports the design and verification of asynchronous process systems." [17] SPIN models are written in PROMELA, and specifications for desired system behavior can be expressed in LTL.

SPIN models generally exhibit non-deterministic behavior. As previously mentioned, they can include non-deterministic PROMELA constructs, and they can also include multiple asynchronous processes whose order of execution is non-deterministic. Thus, SPIN can be used in debugging mode to interactively or randomly simulate one possible execution of the system. SPIN can also be used in verification mode to simulate all possible executions in order to check that the system always meets the desired specifications. For a good reference on LTL and model checking, see Ref. [2]; for a concise PROMELA reference, see Ref. [13]; and the primary SPIN reference is Ref. [18]. In what follows, we provide the minimal background necessary to understand the UAV related models presented in Sect. 4.3.

4.2.1 Linear Temporal Logic

The following definition assumes time passes in discrete steps and that each moment in time has a single successor moment, i.e. that time is linear.

Definition 1. LTL formulas over the set AP of atomic propositions are formed according to the following grammar

$$\varphi ::= \mathbf{true} \mid a \mid \varphi_1 \wedge \varphi_2 \mid \neg\varphi \mid \bigcirc\varphi \mid \varphi_1 \,\mathsf{U}\, \varphi_2$$

where $a \in AP$, "and" $\wedge$ and "not" $\neg$ are the standard Boolean operators, the "next" operator $\bigcirc$ designates that $\bigcirc\varphi$ holds if φ holds at the next time step, and the "until" operator U holds for $\varphi_1 \,\mathsf{U}\, \varphi_2$ if there is some future time at which φ_2 holds, and φ_1 holds at all times until then.

From these basic operators, we can derive other standard Boolean operators such as "or" $\vee$ and "implies" $\rightarrow$ as well as the temporal modalities "eventually" $\Diamond$ and "always" $\Box$. $\Diamond\varphi$ holds if φ holds at some future time, and $\Box\varphi$ holds if φ holds at the current and all future times.

Common LTL properties of interest include *invariance* properties $\Box a$, *liveness* properties $\Diamond a$, *response* properties $\Box(a \rightarrow \Diamond b)$, *impartiality* or *unconditional fairness* properties $\Box\Diamond a$, *justice* or *weak fairness* properties $\Diamond\Box a \rightarrow \Box\Diamond b$, and *compassion* or *strong fairness* properties $\Box\Diamond a \rightarrow \Box\Diamond b$. The last three properties are often used ensure that a system avoids deadlock, e.g. when multiple processes must share time on a single processor. In this case, unconditional fairness ensures that all processes get access to the processor infinitely often, even if they do not request it. Weak fairness ensures that any process that eventually reaches a point at which it always requests access to the processor gets access infinitely often. And strong fairness ensures that every process that requests access infinitely often gets access infinitely often. More examples will be given in Sect. 4.3.

4.2.2 PROMELA

PROMELA models contain three types of objects: variables, processes, and channels. Variables hold data, processes manipulate data stored in variables, and channels send data between processes. PROMELA models also contain different types of operators, functions, control-flow constructs, and macro definitions.

Simulation of a PROMELA model in SPIN involves the discrete execution of process statements in the model. Every time a statement is executed, the *system state* of the model changes. The system state consists of the values of global and local variables, the contents of channel buffers, and the value of every process counter in the system. The system state thus determines which statements in the model are currently enabled, i.e. capable of being executed. Often, multiple statements are enabled, e.g. because more than one process is running or because a process has reached a set of non-deterministic statements. When more than one statement is enabled, any one of the enabled statements can be executed. In interactive debugging mode, the user decides which statement to execute next; otherwise, the simulator makes a random choice. In verification mode, all possible execution paths are exhaustively explored.

The following subsections give a minimal overview of PROMELA. The first subsection discusses the concepts of executability and enabledness. The next three subsections describe variables, processes, and channels. The three subsections after that discuss control-flow constructs, operators and functions, and how to define types and macros. The final subsection describes some useful user-defined types and macros used in Sect. 4.3.

4.2.2.1 Executability and Enabledness

In a PROMELA model, active processes execute asynchronously. Each active process has a process counter that indicates the process's current point of execution, i.e. which statement or statements are currently available. An available statement can only *execute* if it is *enabled*; otherwise, it is *blocked*. Assignments to a variable are always enabled, and expressions are enabled only if they do not evaluate to 0. For instance, if a=1 and b=2, then $a < b$ is enabled and $a > b$ is blocked. If all available statements in a process are blocked, execution of the process halts until one or more of the available statements becomes enabled, e.g. because another process changes the value of a global variable.

Processes execute in an interleaved manner. That is, when more than one active process has enabled statements, any one of the processes may execute an enabled statement. In interactive mode, the user decides which process and statement executes next; otherwise, the simulator decides. In verification mode, all possible interleavings are systematically explored.

4.2.2.2 Variables

Variables and arrays of variables can be declared in PROMELA using the keywords **bit**, **bool**, **byte**, **pid**, **short**, **int**, or **unsigned**. A constant value can be assigned in the declaration if desired. Examples are shown in Listing 4.1. Variables that are declared within a process are *local*; they cannot be accessed by other processes. Variables declared outside of any process are *global*; they can be accessed by any process.

Listing 4.1 Examples of variable declarations and definitions in PROMELA

```
bool flag1, flag2=1, flag3;
int counter;
short a=15;
byte b[10]=1;
```

The domain of integer values a variable can assume depends upon its type and the architecture of the system being used to compile and simulate the model. A **bit** and **bool** are both 1-bit values. Typically, a **byte** and **pid** are both 8-bit unsigned values, a **short** is a signed 16-bit value, an **int** is a signed 32-bit value, and an **unsigned** is an unsigned 16-bit value. An **unsigned** variable x can be restricted to n bits, as shown in Listing 4.2. Restricting the size of variables as much as possible reduces the amount of memory used by SPIN during verification.

Listing 4.2 An example of an unsigned variable declaration

```
unsigned x : n;
```

PROMELA also supports two kinds of user-defined data types. First, a single set of up to 256 global symbolic constants can be defined using the **mtype** keyword. This is similar to a C-style enumeration. Second, the **typedef** keyword can be used to declare a data structure type. See Ref. [13] or Ref. [18] for further details or Sect. 4.2.2.8 for some examples.

4.2.2.3 Processes

Processes are declared and defined using the **proctype** keyword. For example, a process pname() taking **byte** arguments a and b and an **int** argument c as inputs can be declared and defined as shown in Listing 4.3. A process can be instantiated from within another active process using the **run** keyword, as shown in Listing 4.4. Alternatively, processes can be declared such that N instances are active in the initial system state using the **active** keyword, as in Listing 4.5. All N instances are instantiated using default values for the input argument types. In this case, a=b=c=0. Note that PROMELA defines a special process called **init**, which is always active in the initial system state. By convention, **init** is used to initialize global variables and instantiate other processes.

Listing 4.3 Example of a process declaration and definition

```
proctype pname(byte a, b; int c) {
   <PROMELA statements that define the process>
}
```

Listing 4.4 Instantiating a previously declared and defined process

```
run pname(4, 3, 25);
```

Listing 4.5 Instantiating, declaring, and defining N active processes

```
active [N] proctype pname(byte a, b; int c) {
   <PROMELA statements that define the process>
}
```

4.2.2.4 Channels

Channels are declared using the **chan** keyword. Consider the channel declarations in Listing 4.6. The first channel, Input, is one that has been declared but not initialized; an initialized channel must be assigned to Input before Input can be used. The second channel, Dataset, is a single asynchronous channel that can buffer up to two messages, each containing a **bit** and an **int**. Devices is an array of three channels. These channels are synchronous; their buffer sizes are set to 0, and they cannot store messages. Send and receive commands on these channels must be enabled simultaneously in order for data to be exchanged. Thus, send and receive commands on a synchronous channel are blocking, i.e. a send command is not enabled until a corresponding receive command is also enabled and vice versa. Send commands to a full channel are also blocking; process execution cannot proceed until the channel buffer is no longer full, e.g. because another process receives a message on the channel. Receive commands on an empty channel are also blocking. Send commands are indicated by the symbol ! and receive commands by the symbol ?, as shown in Listing 4.7.

Listing 4.6 Examples of channel declarations

```
chan Input;
chan Dataset = [2] of {bit, int};
chan Devices[3] = [0] of {byte};
```

Listing 4.7 Examples of send and receive commands

```
/* Send a message with a bit and an int on channel Datasets */
bit flag1 = 1;
Dataset!flag1 ,15;
/* Receive the message on channel Datasets */
bit flag2; int value;
Dataset?flag2 , value
/* Wait to receive a byte on Device[0] and store it in devInput*/
byte devInput;
Devices[0]?devInput;
```

4.2.2.5 Control Flow

The keywords **goto**, **atomic**, **d_step**, **if**, and **do** are used to control execution flow in a PROMELA model. The **goto** keyword is used in conjunction with a *label* to move a process counter. For instance, in Listing 4.8, the statement **goto** section1 causes execution of the process to jump to the line after the label section1.

Listing 4.8 Example use of the **goto** keyword

```
goto section1;
    <PROMELA statements>
section1:
    <PROMELA statements>
```

An **atomic** statement can be used to restrict the interleaving of processes. The code in Listing 4.9 will attempt to consecutively execute all statements in the **atomic** block if the first statement in the block is enabled. Interleaving execution with other processes is only permitted if a blocked statement is encountered. A **d_step** block is similar, except that interleaving with other processes is never allowed, even if a blocked statement is encountered. Also, execution in a **d_step** block must be deterministic; given the same system state at the start of the block, the system state at the end of the block should always be the same. Furthermore, **goto** statements are not allowed in a **d_step** block. The advantage of **d_step** over **atomic** is a smaller overall system state; because its execution is deterministic, intermediate states representing individual statements in the **d_step** block need not be stored in the system state.

Listing 4.9 Form of an atomic block

```
atomic {
    <PROMELA statements>
}
```

The **if** selection construct can be used to randomly select one of several enabled *options* for execution, as shown in Listing 4.10. Each option consists of a sequence of one or more statements, the first of which is a *guard*. An option is enabled only if its corresponding guard is enabled. If no options are enabled, the **if** construct is blocking. The keyword **else** can be used once in an **if** construct and is enabled only if all other options are blocked. Consider the example in Listing 4.11. If a < 15, the first statement is enabled and a can be incremented. If b < 10, the second statement is enabled and b can be incremented. If both conditions are true, then either a or b can be incremented. If neither condition is true, both a and b are decremented.

Listing 4.10 Form of an **if** statement

```
if
    :: <option 1>
    :: <option 2>
    ...
    :: <option n>
fi;
```

Listing 4.11 An example **if** statement

```
/* An 'if' construct example */
if
    :: (a < 15) -> a++;
    :: (b < 10) -> b++;
    :: else -> a--; b--;
fi;
```

The **do** repetition construct shown in Listing 4.12 is similar to the **if** selection construct, except that it repeatedly selects from among its options until a **break** statement is encountered or a **goto** statement points to a label outside the **do** construct. Consider the example in Listing 4.13. This example increments a and b zero or more times each, as long as a and b are both less than 14. Note that **skip** is a "blank" statement that is always enabled.

Listing 4.12 Form of a **do** statement

```
do
    :: <option 1>
    :: <option 2>
    ...
    :: <option n>
od;
```

Listing 4.13 An example **do** statement

```
/* A 'do' construct example*/
do
    :: (a < 14 && b < 14) -> a++;
    :: (a < 14 && b < 14) -> b++;
    :: skip -> break;
od;
```

4.2.2.6 Operators and Functions

Additional operators and functions that can be used to build expressions in PROMELA are shown in Listing 4.14. The first three rows of operators have the same meaning as in C. In the last row, the function **len**() can be used to check the number of messages stored in a channel's buffer. The functions **empty**(), **nempty**(), **full** (), and **nfull** () can be used to check whether a channel's buffer is empty, not empty, full, or not full. Due to implementation issues, the negation operator ! cannot be used with any of these four functions. The **eval** () function evaluates an expression and turns it into a constant, and the **enabled**() function checks to see if a statement is enabled. The **assert** () statement is used to check simple safety properties. If the expression inside the **assert** () function evaluates to *true*, **assert** () does nothing; if it evaluates to *false*, execution stops and SPIN reports an error.

Listing 4.14 Additional operators and functions in PROMELA

```
+       -        *         /        %       >       >=
<       <=       ==        !=       &       &&      |          ||
!       ^        ~         >>       <<      ++      --
len()   empty()  nempty()  nfull()  full()  eval()  enabled()  assert()
```

4.2.2.7 Types and Macros

Macros that can be processed by a C preprocessor can also be used in a PROMELA model. For instance, the statements in Listing 4.15 define MAXB as a macro for the constant 255 and IMPLIES() with parameters p and q as a macro function for checking the logical implication $p \rightarrow q$.

Listing 4.15 Examples of C-style macros in PROMELA

```
#define MAXB 255;
#define IMPLIES(p,q) ((!p) || (q))
```

PROMELA also supports the **inline** keyword for more complex macros that use PROMELA statements. For example, the **inline** macro function LookUp() in Listing 4.16 can be used to set the value of b based on the value of a.

Listing 4.16 An example of an inline macro in PROMELA

```
inline LookUp(a, b) {
      if
         :: 0  <= a && a <= 5   -> b = 0;
         :: 5  <  a && a <= 10 -> b = 1;
         :: 10 <  a && a <= 20 -> b = 2;
         :: else -> b = 3;
      fi;
}
```

4.2.2.8 Useful User-Defined Types and Macros

Several sets of user-defined types and macros are used to develop the models described in Sect. 4.3. The first is a library of macros for a *bit vector*, implemented as an unsigned piece of memory in which each bit can be set and tested individually. The second is a library of macros for defining a first-in-first-out *queue* of arbitrary size and type. Both libraries are inspired by Ref. [35].

The bit vector library consists of the macros shown in Listing 4.17. The first macro can be used to declare a bit vector of size n named according to the string stored in x. The next macro defines a constant equal to $2^{31}-1$, i.e. a 0-bit followed by 31 1-bits. This is used later in the macro that sets all bits of a vector to 1. The two macros after that can be used to set bit i of bit vector bv to 0 or 1, the next two macros set all bits to 0 or 1, and the last two macros test whether the value of bit i is 0 or 1.

Bit vectors can be easily tested for two other conditions. First, the statement bv == 0 can be used to check whether all bits in bit vector bv are 0. Second, the statement bv > 0 can be used to check whether any of the bits in bv are non-zero.

Listing 4.17 Macros that comprise the bit vector library

```
#define BITV_U(x,n)          unsigned x:n

#define ALL_1S               2147483647

#define SET_0(bv,i)          bv=bv&(~(1<<i))
#define SET_1(bv,i)          bv=bv|(1<<i)
#define SET_ALL_0(bv)        bv=0
#define SET_ALL_1(bv,n)      bv=ALL_1S>>(31-n)

#define IS_0(bv,i)           (!(bv&(1<<i)))
#define IS_1(bv,i)           (!(!(bv&(1<<i))))
```

The queue library consists of the type definitions and macros shown in Listing 4.18. A queue type is declared using the DECLARE_QUEUE(type,size) macro, e.g. DECLARE_QUEUE(**byte**,8). A queue or array of queues can then be declared using the QUEUE(type,size) macro, e.g. QUEUE(**byte**,8) queue1 or QUEUE(**byte**,8) queueArray[5]. ENQUEUE(q,x) queues an element x in queue q, and DEQUEUE(q,x) dequeues the oldest element from q and stores it in variable x. EMPTY(q) removes all elements from queue q, SIZE(q) returns the current number of elements in q, and IS_EMPTY(q) tests whether or not q is empty.

Listing 4.18 Macros and type definitions that comprise the queue library

```
/* Type definition */
#define DECLARE_QUEUE(T,MAX) \
typedef Queue_##T##MAX { \
  T       queue[MAX]; \
  byte    length; \
  byte    ii; \
}

/* Constructor */
#define QUEUE(T,MAX) Queue_##T##MAX

/* Enqueue operation */
inline ENQUEUE(q,x) {
  d_step {
    q.queue[q.length]=x;
    q.length++;
  }
}

/* Dequeue operation */
inline DEQUEUE(q,x) {
  d_step {
    x = q.queue[0];
    q.ii = 1;
    do
      :: (q.ii < q.length)  -> q.queue[q.ii-1] = q.queue[q.ii];
                               q.ii++;
      :: (q.ii == q.length) -> q.queue[q.ii-1] = 0; q.ii++;
      :: (q.ii > q.length)  -> q.length--; break;
    od;
    q.ii=0;
  }
}

/* Dequeue operation */
inline EMPTY(q) {
  d_step {
    do
      :: (q.ii > 0)  -> q.queue[q.ii-1] = 0; q.ii--;
      :: (q.ii == 0)  -> q.length = 0; break;
    od;
  }
}

#define SIZE(q)        (q.length)
#define IS_EMPTY(q)    (q.length == 0)
```

4.2.3 SPIN

The SPIN model checker can either be used to debug a PROMELA model or to exhaustively verify that a PROMELA model's behavior conforms to a set of desired specifications. Debugging involves simulating a small number of possible executions of the model either interactively or randomly. Debugging is mainly used to check for modeling errors. Once the model has been debugged, SPIN can be used in verification mode to check whether all possible executions of the model conform to the desired specifications. If not, SPIN provides a counterexample in the form of an execution trace. This execution trace demonstrates how the model is able to violate one of the specifications. If the model is very large, a partial verification can be performed using a technique known as *bitstate hashing*.

Verification consists of using SPIN to create a verification model in C, which can then be compiled and run. For a model model.promela, SPIN is run at the command line using one of the commands shown in Listing 4.19. The executables spin and pan.exe both have additional options. See Ref. [18] for more information.

Listing 4.19 Commands used to run SPIN and the verification model pan.exe

```
spin model.promela    # run a random simulation
spin -i model.promela # run an interactive simulation
spin -a model.promela # generate a verification model pan.c
                      # from model.promela
gcc -o pan.exe pan.c  # compile the verification model pan.c
./pan.exe -a          # run the verification model
./pan.exe -a -f       # run the verification model, interleaving
                      # processes using weak fairness
```

Specifications can be written in LTL, described in Sect. 4.2.1. In SPIN, the LTL operators $\square$, $\Diamond$, U, $\wedge$, $\vee$, $\neg$, and $\rightarrow$ are written as [], <>, U, &&, ||, !, and ->, respectively. LTL specifications can be given in a PROMELA model using the **ltl** keyword; they take the form **ltl** [propertyName] '{'LTL formula'}'. Examples are shown in Listing 4.20.

Listing 4.20 Example LTL specifications in PROMELA and SPIN

```
ltl response {[](a -> <>b)}
ltl justice {<>([]a) -> [](<>b)}
ltl compassion {[](<>a) -> [](<>b)}
```

4.3 UAV Applications

The following applications show how model checking can be used in the design and verification of multi-agent autonomous systems, especially those containing UAVs. They demonstrate the types of desired properties that can be expressed, the level of abstraction at which the designs are created, and how various types of behavior can be modeled. Each application has different characteristics and requirements, resulting in an exploration of many of the different features and capabilities of model checkers and temporal logic, specifically those found in the SPIN model checker [17], the modeling language PROMELA [13], and the temporal logic LTL.

The first application involves high-level design and verification of a UAV cooperator controller for a ground sensor monitoring task. This application relies on a centralized scheme for coordinating the actions of multiple UAVs that must continually visit a set of unattended ground sensors in order to collect data and aggregate it at a centralized location. This application is discussed in Sect. 4.3.1. The second application involves high-level design of a leader election protocol for a network of unattended ground sensors. In this application, unattended ground sensors individually collect data related to the position of a mobile intruder. The data is sent to a pursuing UAV. However, due to bandwidth restrictions, it is desirable to have the sensors elect a leader to aggregate the data before it is sent to the UAV. This application is discussed in Sect. 4.3.2. The third application involves verification of high-level UAV mission plans in an escort scenario. Plans could be generated by either a human, automation, or a combination of both. This application is discussed in Sect. 4.3.3.

4.3.1 A Centralized UAV Cooperative Controller for a Continuous Sensor Monitoring Task

In this application, the goal is to develop a multi-UAV cooperative control system for continuous monitoring of a set of unattended ground sensors (UGSs). Specifically, the controller should create UAV patrol routes such that every active UGS is eventually visited by at least one active UAV. Each UAV gathers data from the UGSs on its patrol route and takes the collected data back to a forward operating base, as depicted in Figure 4.1.

The system consists of one or more UAVs, one or more UGSs, and a forward operating base. A UAV and UGS can only communicate if their communication ranges overlap, e.g. over a Wi-Fi connection with limited range. For simplicity, we assume that a UAV must fly directly over a UGS in order to communicate with it. A UAV may change status from active to inactive when it returns to the base, e.g. to land for refueling. A UGS may also change status from active to inactive, e.g. if it is found to be damaged or unresponsive. UAVs and UGSs may also change status from inactive to active, e.g. after being refueled or repaired.

Assume there are $m > 1$ UAVs, $n \geq m$ UGSs, one forward operating base, and that the system operates over an infinite time horizon. The forward operating base acts as a centralized cooperative controller that assigns patrol routes to active UAVs. Whenever a UAV or UGS changes status from active to inactive or vice versa, the controller must replan the patrol routes. For simplicity, assume that these status changes are only registered when a UAV returns to the base, and a UAV only starts a new patrol route after completing its previous patrol route or after being reactivated.

Let $ugsRequest_i$ and $ugsVisited_i$ for $i \in \{0,\dots,n-1\}$ be boolean variables; labeling starts at 0 because PROMELA uses 0-based indexing. Dependence on time is

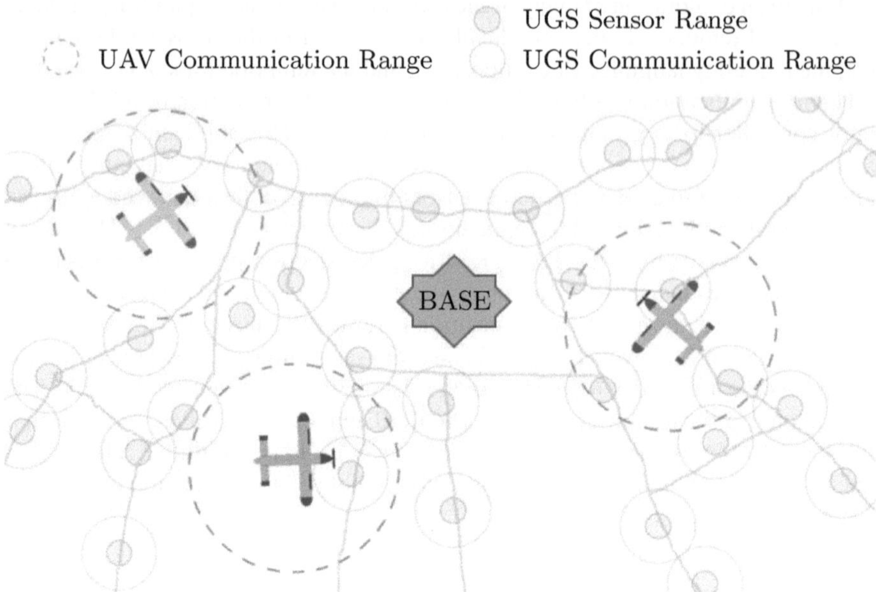

Fig. 4.1 The basis for the UAV cooperative control scenario. The goal in this scenario is to have active UAVs fly over active UGSs, collect data from them, and return the data to the base. UAVs and UGSs may change status from active to inactive and vice versa throughout the scenario.

implicit; these variables can change as the system changes state. Let $ugsRequest_i$ be **true** when a visit request is being made for active UGS i, and let $ugsVisited_i$ be **true** when UGS i is being visited by an active UAV. At all other times, these propositions are **false**. Visit requests are generated for all active UGSs whenever the base replans patrol routes, and they are generated for specific UGSs on a patrol route when a UAV repeats that route. UGS visits are registered when a UAV communicates with a UGS. Assume at least one UAV is always active; otherwise, there is no way to visit the UGSs.

Desired properties for the cooperative controller can be expressed in LTL as:

1. $\Box(ugsRequest_i \rightarrow \Diamond ugsVisited_i)$ for $i = 0, \ldots, n-1$
2. $\Diamond\Box ugsRequest_i \rightarrow \Box\Diamond ugsVisited_i$ for $i = 0, \ldots, n-1$
3. $\Box\Diamond ugsRequest_i \rightarrow \Box\Diamond ugsVisited_i$ for $i = 0, \ldots, n-1$

Property 1 is a response property. Every time a request to visit a UGS generated, the UGS should eventually be visited. Property 2 is a weak fairness property. If there is eventually always a request to visit a UGS, then the UGS should be visited infinitely often. Property 3 is a strong fairness property. If requests to visit a UGS are generated infinitely often, then the UGS should be visited infinitely often.

4.3.1.1 The Model

The PROMELA model of the system consists of two process definitions: Base(), which models the behavior of the forward operating base, and Uav(), which models the behavior of a UAV. To keep the system space small, the model assumes a maximum of two UAVs and four UGSs. This allows checking cases in which only one UAV is active, multiple UAVs are active, multiple UAVs are active and then a UAV goes offline, and a single UAV is active and then another comes online. It also allows checking similar cases for the UGSs.

Using the macros described in Sect. 4.2.2.8, the model declares the global bit vectors in Listing 4.21 to hold the values of $ugsRequest_i$ and $ugsVisited_i$ for $i = 0,\ldots,3$. Then, some of the properties to be verified with respect to UGS 0 are shown in Listing 4.22. Similar properties can be defined and checked for the other three UGSs.

Listing 4.21 Bit vectors used to track UGS requests and visits in the UAV cooperative control scenario

```
/* Global bit vectors for UGS requests and visits */
BITV_U(ugsVisited,4); BITV_U(ugsRequest,4);
```

Listing 4.22 LTL properties to be verified in the UAV cooperative control scenario

```
ltl response0 {[](IS_1(ugsRequest,0) -> <>IS_1(ugsVisited,0))}
ltl justice0 {<>[]IS_1(ugsRequest,0) -> []<>IS_1(ugsVisited,0)}
ltl compassion0 {[]<>IS_1(ugsRequest,0) -> []<>IS_1(ugsVisited,0)}
```

The Base() process must track which UAVs and UGSs are active, manage UAV and UGS assignments, and communicate these assignments to the UAVs. For simplicity, two separate, asynchronous channels are used to communicate UGS assignments to the UAVs, as shown in Listing 4.23. To facilitate UGS activation and assignment, the model includes a function macro AssignUGSs(). This function randomly activates at least as many UGSs as active UAVs and randomly assigns them to active UAV patrol routes. See Listing 4.51 in the Appendix.

Listing 4.23 Channels used by the base to send assignments to the UAVs

```
chan baseUAV0 = [1] of {byte};
chan baseUAV1 = [1] of {byte};
```

The behavior of Base() is shown in Figure 4.2. The PROMELA code that implements this process is shown in Listing 4.24. It contains two major sections labeled assignUGSs and monitorUavs. The assignUGSs section uses the inline macro AssignUGSs() to randomly activate and assign UGSs to UAVs. It also triggers visit requests for active UGSs and sends UGS patrol assignments to the UAVs. If a UAV is inactive, it will receive a 0-bit vector for its assignments. The monitorUavs section then waits until all UAVs have received their new assignments. It then determines whether or not a UAV will change status from active to inactive or vice versa on

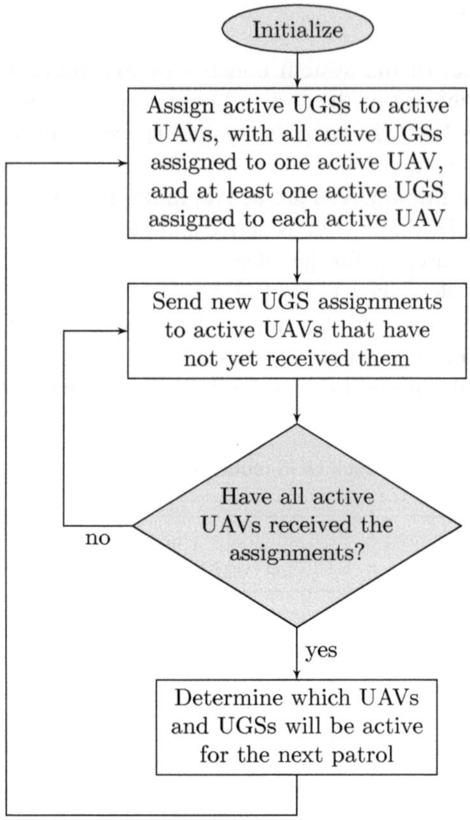

Fig. 4.2 A flowchart describing behavior of the Base() process in the UAV cooperative control scenario

its next patrol, with the constraint that at least one UAV must always be active. The Base() process then returns to the section labeled assignUGSs and the process repeats.

The process Uav() is used to model each of the UAVs. Its behavior is shown in Figure 4.3, and the code that implements it is shown in Listing 4.25. Uav() takes two input arguments: a channel and a boolean value. The former is used to receive UGS assignments; the latter is for identification purposes. At the label returnToBase, Uav() receives new assignments from Base() over the channel if any are available. If not, it triggers a request to visit the UGSs previously assigned to it. Next, Uav() simulates visiting the assigned UGSs in a random order. It then jumps back to the label returnToBase and repeats. Recall that if the Base() process determines that the UAV should be inactive, the vector of assignments is 0. In that case, Uav() will not trigger any UGS visit requests or UGS visits because all the bits in ugsList are 0.

Listing 4.24 The Base() process

```
active proctype Base() {
  /* Bit vectors indicating active UGSs/UAVs */
  BITV_U(ugsActive,4); BITV_U(uavActive,2);
  SET_1(uavActive,0);  SET_1(uavActive,1);

  /* Bit vectors to store UGSs for UAVs 0 and 1 to visit */
  BITV_U(ugsList0,4); BITV_U(ugsList1,4);

  /* Start running the UAVs */
  run Uav(baseUAV0,0); run Uav(baseUAV1,1);

assignUGSs:
  /* Determine which UGSs should be active. Set UGS assignments */
  AssignUGSs(uavActive,ugsActive,ugsList0,ugsList1);

  /* Trigger UGS requests */
  ugsRequest = ugsActive;
  SET_ALL_0(ugsRequest);

  /* Send UGS assignments through the channels */
  atomic{ baseUAV0!ugsListUav0; baseUAV1!ugsListUav1; }

monitorUavs:
  if /* Make sure each UAV gets its assignment */
    :: (empty(baseUAV0) && empty(baseUAV1)) -> skip;
    :: (nempty(baseUAV0) || nempty(baseUAV1)) -> goto monitorUavs;
  fi;

  if /* Once a patrol is finished, UAV might go on or offline */
    :: IS_1(uavActive,0) && IS_1(uavActive,1)->SET_0(uavActive,0);
    :: IS_1(uavActive,0) && IS_1(uavActive,1)->SET_0(uavActive,1);
    :: IS_0(uavActive,0)->atomic{SET_1(uavActive,0);
                                 SET_0(uavActive,1);}
    :: IS_0(uavActive,1)->atomic{SET_1(uavActive,1);
                                 SET_0(uavActive,0);}
    :: skip;
  fi;

  goto assignUGSs;
}
```

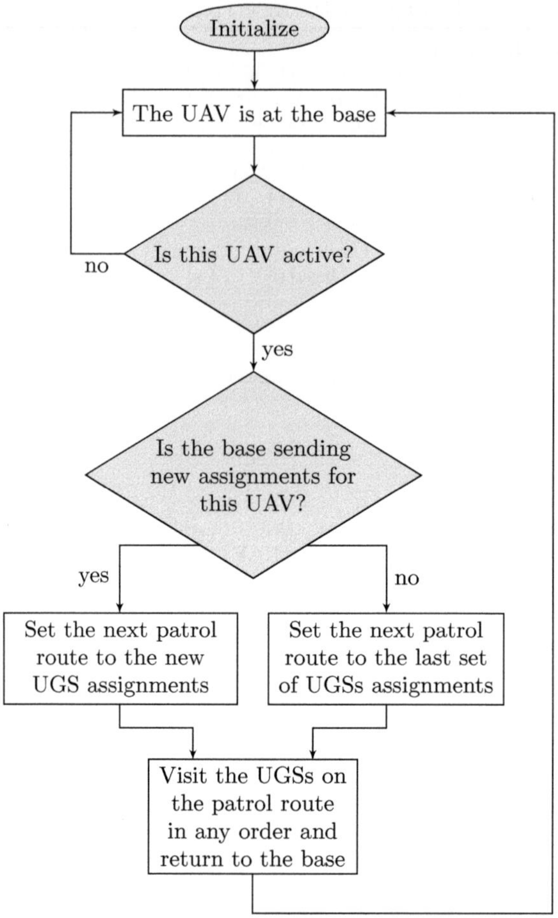

Fig. 4.3 A flowchart describing behavior of the Uav() process in the UAV cooperative control scenario

Listing 4.25 The Uav() process

```
proctype Uav(chan channel; bool uavNum) {
  /* Declare a bit vector indicating which UGSs to visit */
  BITV_U(ugsList,4);

returnToBase:
  /* If the base sends a new assignments, update the list.
     Otherwise, resignal requests for the last set of UGSs. */
  if
    :: nempty(channel)->channel?ugsList;
    :: empty(channel)->ugsRequest=ugsList; SET_ALL_0(ugsRequest);
  fi;

  /* Make a copy of ugsList to be modified */
  ugsToVisit = ugsList;

  /* Visit the assigned UGSs in random order */
  BITV_U(ugsToVisit,4);
  do
    :: IS_1(ugsToVisit,0)->SET_1(ugsVisited,0);
                           SET_0(ugsVisited,0);
                           SET_0(ugsToVisit,0);
    :: IS_1(ugsToVisit,1)->SET_1(ugsVisited,1);
                           SET_0(ugsVisited,1);
                           SET_0(ugsToVisit,1);
    :: IS_1(ugsToVisit,2)->SET_1(ugsVisited,2);
                           SET_0(ugsVisited,2);
                           SET_0(ugsToVisit,2);
    :: IS_1(ugsToVisit,3)->SET_1(ugsVisited,3);
                           SET_0(ugsVisited,3);
                           SET_0(ugsToVisit,3);
    :: else -> break;
  od;

  goto returnToBase;
}
```

4.3.1.2 Verification

The model – named UavUgsControl.promela – can now be put through the verification process. For instance, the commands in Listing 4.26 can be executed in a UNIX terminal to check the property response0. In this case, the verifier reports one error: an *acceptance cycle* is found. This indicates that the property response0 is not satisfied [17, 18]. An error trail that demonstrates how the property is violated can be generated using the –t and –p flags, as shown in Listing 4.27.

Listing 4.26 Commands to verify the response0 property

```
spin -a UavUgsControl.promela # build verification model pan.c
gcc -o pan.exe pan.c          # compile pan.c
./pan.exe -a -N response0     # check property response0
```

Listing 4.27 Printing the error trail

```
spin -t -p UavUgsControl.promela # show error trail
```

Listing 4.28 shows the statements that are executed at the end of the trail. The trail reveals that after sending the UAVs their assignments, the Base() process continually executes the statement (**nempty**(baseUAV0) || **nempty**(baseUAV1)) −> **goto** monitorUavPatrols. Thus, the UAVs never receive and remove their assignments from the channels. This is because, by default, SPIN does not require that processes execute fairly; the Base() process can continually execute commands without ever allowing a Uav() process to execute. However, this is not realistic behavior for our model. In a real scenario, the UAVs and forward operating base will all regularly perform actions. We can simulate this behavior by instructing SPIN to use weak fairness when executing processes. That is, any process that is continuously enabled will execute infinitely often. This is done in the verifier using the −f flag, as in Listing 4.29. Using weak fairness, all LTL properties are successfully verified.

Listing 4.28 Checking the error trail

```
...
14: proc 0 (Base) ... :140 (state 71)    [ugsRequest = ugsActive]
16: proc 0 (Base) ... :141 (state 72)    [ugsRequest = 0]
18: proc 0 (Base) ... :145 (state 73)    [baseUAV0!ugsListUav0]
19: proc 0 (Base) ... :145 (state 74)    [baseUAV1!ugsListUav1]
<<<<<START OF CYCLE>>>>>
21: proc 0 (Base) ... :151 (state 78)   [((nempty(baseUAV0) ||
                                           nempty(baseUAV1)))]
```

Listing 4.29 Verifying the UAV cooperative controller under weak fairness conditions

```
./pan.exe -a -f -N response0
```

It should be noted that some changes were required in the model in order to enable weak fairness. Weak fairness requires that all processes eventually reach a point at which they are always enabled; thus, statements that can block should be avoided. Consider the channel monitoring commands in section monitorUavs of process Base(), shown in Listing 4.30. Since line 4 simply loops back to the **if** statement, we could normally replace all 5 lines with the statement **empty**(baseUAV0) && **empty**(baseUAV1). This would prevent the Base() from generating new assignments until each of the UAVs received the last set of assignments, as desired. However, weak fairness would no longer apply, since the statement would block when a channel is full. Thus, we explicitly include the case in which a channel is full in the **if** statement, ensuring that there is always an executable statement. Thus, weak fairness can be enforced.

Listing 4.30 Channel monitoring in the Base() process

```
monitorUavs:
  if /* Ensure each UAV gets assignments so it can do a patrol */
    :: (empty(baseUAV0) && empty(baseUAV1)) -> skip;
    :: (nempty(baseUAV0) || nempty(baseUAV1)) -> goto monitorUavs;
  fi;
```

Verification also revealed a bug in an earlier version of the model. Originally, the returnToBase section of the Uav() process contained the statements shown in Listing 4.31. These statements avoid going through the rest of the process if ugsList is 0. When a new assignment is received, it is stored in ugsList. On the next iteration of the **do** statement, the channel should be empty, and the process can break out of the **do** statement if ugsList > 0. However, a problem was revealed during verification. It is possible for the Uav() process to receive one assignment, the Base() process to generate another assignment, and the Uav() process to receive the next assignment without completing the first. Thus, requests to visit a UGS can go unanswered. This section was changed in the final version of the model to fix the error.

Listing 4.31 Original code used in Uav()

```
  /* Declare a bool to indicate a new assignment was received */
  bool newAssignment;

returnToBase:
  /* If the base sends new assignments, update the UGS list.
     Otherwise, resignal requests for the last set of UGSs. */
  newAssignment = 0;
  do
    :: nempty(channel) -> channel?ugsList; newAssignment = 1;
    :: empty(channel) && ugsList > 0 -> break;
    :: empty(channel) && ugsList == 0 -> goto returnToBase;
  od;
  if
   :: !newAssignment -> ugsRequest=ugsList; SET_ALL_0(ugsRequest);
   :: else -> skip;
  fi;
```

4.3.2 A Decentralized Leader Election Protocol for a Network of Unattended Ground Sensors

The goal in this application is to design a leader election protocol for a network of unattended ground sensors that autonomously detect an intruder and send estimates of its location to a pursuing UAV. The scenario is depicted in Figure 4.4 and is similar to the one described in [38]. The ground sensors are capable of detecting and measuring local disturbances, e.g. using a magnetometer, and sending readings of these measurements to other sensors over a wireless network. Accurate estimation of the intruder's position requires readings from multiple sensors. However, the bandwidth of the network is assumed to be low. Thus, it is preferable to have only a small number of sensors sending information to the UAV. One solution is to elect a leader, i.e. a single sensor that aggregates measurements from all other sensors, estimates the position of the intruder, and sends this information to the pursuer. However, a centralized leader election protocol is expensive in terms of bandwidth and is susceptible to network outages and delays. Thus, a decentralized election scheme is more realistic.

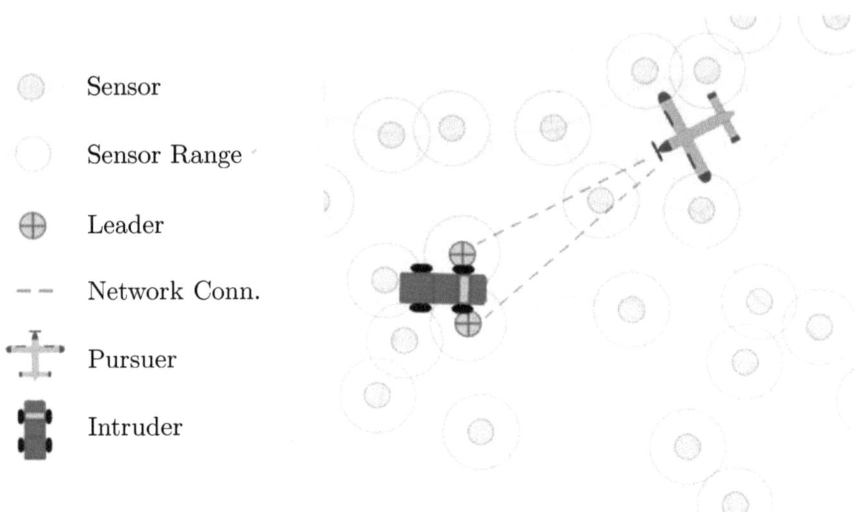

Fig. 4.4 The basis for the unattended ground sensor scenario. Networked ground sensors detect the presence of an intruder when it is within range and broadcast measurements to other sensors in the network. When a sensor detects that it currently has the highest measurement, it elects itself leader. Leaders aggregate measurements and send estimates of the intruder's position to the pursuing UAV.

In the decentralized scheme presented here, every sensor compares its own measurement to those it has received from others. When a sensor detects that its measurement is currently the highest, it elects itself leader. Because the scheme is decentralized, multiple leaders are allowed. To limit bandwidth usage, sensors must wait at least .5 seconds before broadcasting consecutive readings. Also, a sensor cannot elect itself leader or send estimates to the pursuer more than once every .5 seconds. Sensors act asynchronously, and so this .5 second limit establishes a "weak epoch" for each sensor. Also, all measurements include a timestamp and are considered invalid after .5 seconds.

Assume there are n sensors in the network. Let leader be a bit vector with n elements such that element i is 1 if Sensor i is currently leader and 0 otherwise.

Then, the goal is to design a protocol for the network that satisfies the LTL property $\square(leader > 0)$.

4.3.2.1 The Model

Figure 4.5 shows the two conceptual parallel processes that run on each sensor. Let *myLvl* be the level or value of the measurement currently being registered by the sensor, and let *maxLvl* be the highest non-expired level or measurement received by the sensor. The flowchart on the left shows how the sensor broadcasts measurements. As soon as the sensor detects that $myLvl > 0$, it broadcasts its measurement level to other sensors in the network. Then, it waits .5 seconds, the measurement expires, and the sensor can repeat the broadcasting process. The flowchart on the right shows how the sensor elects itself leader. As soon as the sensor determines that $myLvl \geq maxLvl$, it elects itself leader and sends data to the pursuer. Then, it waits .5 seconds, its leadership status expires, and the sensor can repeat the election process.

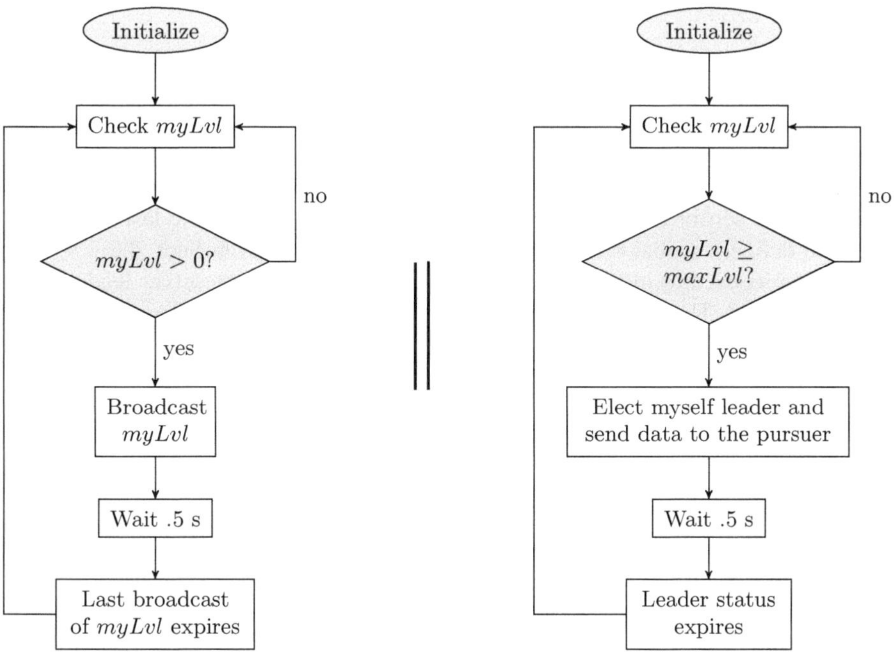

Fig. 4.5 Flowchart depicting the process used to model the behavior of each sensor in the network

Our PROMELA model of the sensor network consists of three sensors. To begin, we define some macros in Listing 4.32 that simplify the design. The first macro defines the number of sensors in the network, the second defines the length of queues needed to track network events, and the third defines the constant value for a three-element bit vector whose entries are all 1. The next set of macros define constants that represent timed events for each sensor: *detection* and broadcasting of a non-zero measurement and self leadership *election*.

Listing 4.32 Macros defined in the model of the ground sensor network

```
/****** Model variables ******/
#define NUM_SENSORS   3
#define QUEUE_LENGTH  6
#define THREE_1S      7

/****** Constants to define time constrained actions ******/
#define DETECTION_0   0
#define ELECTION_0    1
#define DETECTION_1   2
#define ELECTION_1    3
#define DETECTION_2   4
#define ELECTION_2    5
```

Next, in Listing 4.33 we define a global array sensorLvls with three elements used to store sensor measurements that have been broadcast in the last .5 seconds. We also define a global bit vector leader – using macros from Sect. 4.2.2.8 – that has three bits used to indicate which sensors have elected themselves leader in the last .5 seconds. The bits in this vector are initialized to 1, i.e. all sensors start as leaders.

Listing 4.33 A global array and bit vector used to track sensor status

```
/****** Vector of broadcast sensor values ******/
byte sensorLvls[3] = 0;

/****** Bit vector to designate leader statuses ******/
BITV_U(leader,3) = THREE_1S;
```

The bit array leader can then be used to check whether or not there is at least one leader. Thus, the LTL property we want to verify for the model is as given in Listing 4.34.

Listing 4.34 The LTL property to be verified for the sensor network

```
ltl leaderEveryEpoch { [](leader > 0) }
```

Modeling the passage of time in PROMELA can be difficult. But since each sensor must wait at least .5 seconds after a detection or self election, events can be tracked by the order in which they occur. We use the global variables in Listing 4.35 for this purpose.

Listing 4.35 Variables used to track time constrained events

```
/****** Variables used to track time constrained events ******/
BITV_U(canElectSelf,3) = THREE_1S;
BITV_U(canSendDetection,3) = THREE_1S;
DECLARE_QUEUE(byte,8);
QUEUE(byte,8) eventQueue;
```

As events occur, they are placed in a queue called eventQueue. Bit vectors are used to indicate which time constrained events can occur; when an event occurs for a particular sensor, the corresponding bit in the corresponding bit vector is set to 0. An event being dequeued implies that .5 seconds have passed since the event last occurred, and the corresponding bit is set back to 1. The bit vector canElectSelf is used to track leader election events; canSendDetection is used to track sensor detection and broadcast events.

There are two processes defined in the network model. The first is EpochMonitor(), shown in Listing 4.36, which manages eventQueue and makes changes to the model when events expire. It also instantiates three instances of the process SensorDetection(), shown in Listing 4.37. The three instances are numbered 0, 1, and 2. The process SensorDetection() randomly triggers sensor detections of different levels; since there are three nodes in the network, the levels are 1, 2, and 3. This covers the possibility that all three sensors detect different measurement levels. It could also be that two or more sensors register the exact same level.

Listing 4.36 The EpochMonitor() process used to manage timed events

```
active proctype EpochMonitor() {

  byte event; byte ii;

  /* Register initial leader election events */
  atomic{
    ENQUEUE(eventQueue,ELECTION_0); ENQUEUE(eventQueue,ELECTION_1);
    ENQUEUE(eventQueue,ELECTION_2);
  }

  /* Start the sensors */
  run SensorDetection(0); run SensorDetection(1); run SensorDetection(2);

  /* Events in the queue can expire in the order they were queued */
  do
    :: atomic{!IS_EMPTY(eventQueue) ->

       /* Get the oldest event and update the queue */
       DEQUEUE(eventQueue, event);

       /* Make changes depending on what the event was */
       if
         :: event==DETECTION_0 -> sensorLvls[0] = 0; SET_1(canSendDetection,0);
                                  printf("EpochMonitor: DETECTION_0 ending.\n");
         :: event==DETECTION_1 -> sensorLvls[1] = 0; SET_1(canSendDetection,1);
                                  printf("EpochMonitor: DETECTION_1 ending.\n");
         :: event==DETECTION_2 -> sensorLvls[2] = 0; SET_1(canSendDetection,2);
                                  printf("EpochMonitor: DETECTION_2 ending.\n");
         :: event==ELECTION_0  -> SET_1(canElectSelf,0); SET_0(leader,0);
                                  printf("EpochMonitor: ELECTION_0 ending.\n");
         :: event==ELECTION_1  -> SET_1(canElectSelf,1); SET_0(leader,1);
                                  printf("EpochMonitor: ELECTION_1 ending.\n");
         :: event==ELECTION_2  -> SET_1(canElectSelf,2); SET_0(leader,2);
                                  printf("EpochMonitor: ELECTION_2 ending.\n");
      fi;
      printf("EpochMonitor: sensorLvls:[%d,%d,%d], leader*:[%d,%d,%d]\n",
        sensorLvls[0], sensorLvls[1], sensorLvls[2],
        IS_1(leader,0), IS_1(leader,1), IS_1(leader,2));

      /* When an event expires, see if a sensor can now elect itself leader */
      do
         :: (IS_1(canElectSelf,0) && sensorLvls[0] >= sensorLvls[1] &&
              sensorLvls[0] >= sensorLvls[2]) -> SET_0(canElectSelf,0);
              SET_1(leader,0); ENQUEUE(eventQueue,ELECTION_0);
              printf("Sensor 0: Electing self leader.\n");
         :: (IS_1(canElectSelf,1) && sensorLvls[1] >= sensorLvls[0] &&
              sensorLvls[1] >= sensorLvls[2]) -> SET_0(canElectSelf,1);
              SET_1(leader,1); ENQUEUE(eventQueue,ELECTION_1);
              printf("Sensor 1: Electing self leader.\n");
         :: (IS_1(canElectSelf,2) && sensorLvls[2] >= sensorLvls[0] &&
              sensorLvls[2] >= sensorLvls[1]) -> SET_0(canElectSelf,2);
              SET_1(leader,2); ENQUEUE(eventQueue,ELECTION_2);
              printf("Sensor 2: Electing self leader.\n");
         :: else -> break;
     od;

      printf("EpochMonitor: sensorLvls:[%d,%d,%d], leader:[%d,%d,%d]\n",
        sensorLvls[0], sensorLvls[1], sensorLvls[2],
        IS_1(leader,0), IS_1(leader,1), IS_1(leader,2));
    }
  od;
}
```

Listing 4.37 The SensorDetection() process used to generate sensor detection events

```
proctype SensorDetection(byte myNum) {

  byte myLvl;

cycleStart:

  /* The sensor detects a signal greater than 0 */
  atomic{
    if
      :: myLvl = 1;
      :: myLvl = 2;
      :: myLvl = 3;
    fi;

    /* Register the level and log a detection event */
    sensorLvls[myNum] = myLvl; SET_0(canSendDetection,myNum);
    ENQUEUE(eventQueue,2*myNum);
    printf("Sensor %d: Sending detection of %d.\n", myNum, myLvl);

    /* If canElectSelf, see if myLvl is the highest. If so, elect self */
    printf("Sensor %d: Comparing levels sensorLvls: [%d, %d, %d]\n", myNum,
       sensorLvls[0], sensorLvls[1], sensorLvls[2]);
    if
      :: (IS_1(canElectSelf,myNum) && myLvl >= sensorLvls[0] &&
          myLvl >= sensorLvls[1] && myLvl >= sensorLvls[2]) ->
          SET_0(canElectSelf,myNum); SET_1(leader,myNum);
          ENQUEUE(eventQueue,2*myNum+1);
          printf("Sensor %d: Electing self leader.\n", myNum);
      :: (IS_0(canElectSelf,myNum) && myLvl >= sensorLvls[0] &&
          myLvl >= sensorLvls[1] && myLvl >= sensorLvls[2]) ->
          printf("Sensor %d: Level high, but unable to elect self.\n", myNum);
      :: else -> skip;
    fi;
    printf("Sensor %d: leader: [%d, %d, %d]\n", myNum,
       IS_1(leader,0), IS_1(leader,1), IS_1(leader,2));
  }

  /* Wait until I can send another detection */
  IS_1(canSendDetection,myNum);

  /* Repeat the cycle */
  goto cycleStart;
}
```

4.3.2.2 Verification

As in Sect. 4.3.1, SPIN can be used to verify whether or not the model meets the desired specifications. In this case, the specification is $\square(leader > 0)$. First, SPIN can be used to simulate a random execution of the model, as shown in Listing 4.38. In this execution, each of the sensors registers a detection event. Since the sensors all start as leaders, the previous election events must expire before the sensors can elect themselves again. Sensors 0 and 1 have lower measurement levels than Sensor 3; thus, when their election events expire, they are not able to re-elect themselves. When the election event for Sensor 2 expires, leader is temporarily 0. However,

Sensor 2 immediately re-elects itself leader since it has the highest measurement level. Because this happens in the middle of the **atomic** loop in EpochMonitor() and because there are no blocking statements in the loop, this temporary state is not registered. Thus, it does not violate the specification $\square(leader > 0)$. This is okay; the goal is to check that there is a leader every epoch or every .5 seconds. The leadership status change occurs between epochs, and the time between epochs is considered to be nearly instantaneous. The execution fragment then continues, with Sensor 0 registering a detection event with measurement level 3 and electing itself leader. The model appears to be working correctly, but SPIN can be used in verification mode to perform a full verification. For this model, the property $\square(leader > 0)$ is in fact verified.

Listing 4.38 A random execution of the ground sensor network model

```
    Sensor 0: Sending detection of 1.
    Sensor 0: Comparing levels. sensorLvls: [1, 0, 0]
    Sensor 0: Level high, but unable to elect self.
    Sensor 0: leader: [1, 1, 1]
        Sensor 1: Sending detection of 1.
        Sensor 1: Comparing levels. sensorLvls: [1, 1, 0]
        Sensor 1: Level high, but unable to elect self.
        Sensor 1: leader: [1, 1, 1]
            Sensor 2: Sending detection of 3.
            Sensor 2: Comparing levels. sensorLvls: [1, 1, 3]
            Sensor 2: Level high, but unable to elect self.
            Sensor 2: leader: [1, 1, 1]
EpochMonitor: ELECTION_0 ending.
EpochMonitor: sensorLvls: [1, 1, 3], leader*: [0, 1, 1]
EpochMonitor: sensorLvls: [1, 1, 3], leader: [0, 1, 1]
EpochMonitor: ELECTION_1 ending.
EpochMonitor: sensorLvls: [1, 1, 3], leader*: [0, 0, 1]
EpochMonitor: sensorLvls: [1, 1, 3], leader: [0, 0, 1]
EpochMonitor: ELECTION_2 ending.
EpochMonitor: sensorLvls: [1, 1, 3], leader*: [0, 0, 0]
Sensor 2: Electing self leader.
EpochMonitor: sensorLvls: [1, 1, 3], leader: [0, 0, 1]
EpochMonitor: DETECTION_0 ending.
EpochMonitor: sensorLvls: [0, 1, 3], leader*: [0, 0, 1]
EpochMonitor: sensorLvls: [0, 1, 3], leader: [0, 0, 1]
    Sensor 0: Sending detection of 3.
    Sensor 0: Comparing levels. sensorLvls: [3, 1, 3]
    Sensor 0: Electing self leader.
    Sensor 0: leader: [1, 0, 1]
```

4.3.3 UAV Mission Planning in an Escort Scenario

The goal in this application is to perform model checking on mission plans generated during a VIP escort scenario. In this simple scenario, a ground-based vehicle referred to as a "very important person" or "VIP" must safely traverse a road network in order to reach a desired destination. The VIP is escorted by several UAVs, each of which is able to categorize road segments as safe or unsafe. A specific instance of this scenario is shown in Figure 4.6.

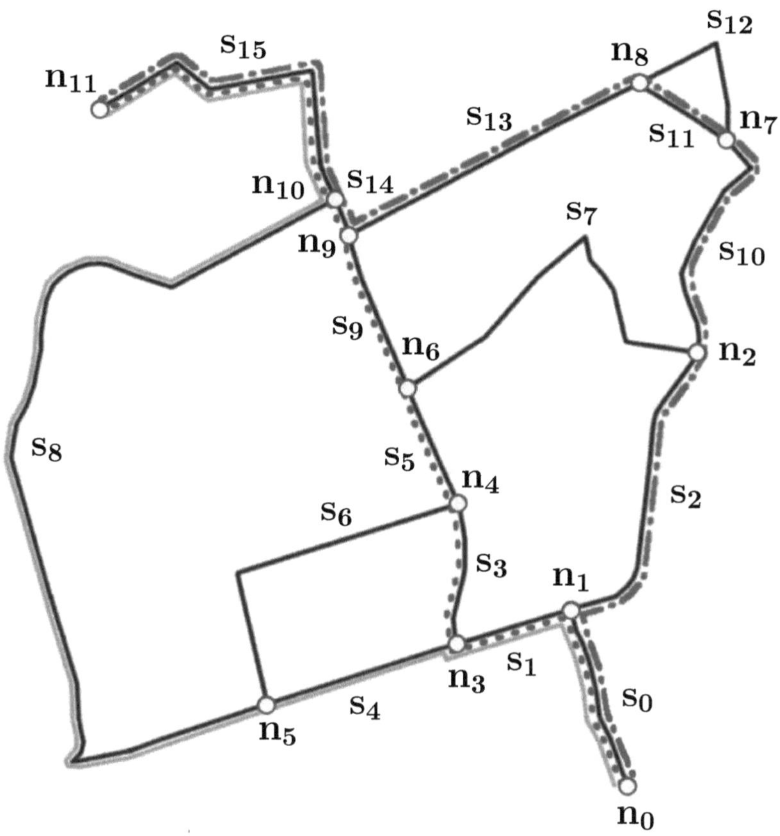

Fig. 4.6 The basis for VIP escort scenario. A ground-based "VIP" vehicle starts at node n_0 and must move safely to node n_{11}. Three UAVs are available to scan road segments to ensure the path travelled by the VIP is safe. The VIP should never traverse a road segment that is unsafe.

Assume there are three UAVs in the scenario. Let $uavLoc_i$ for $i = \{1,2,3\}$ equal the number of the segment currently being scanned by UAV i, and let $vipLoc$ equal the number of the segment currently being travelled by the VIP. Let $safeS_j$ for $j = \{1,2,\ldots,15\}$ be a set of Boolean variables denoting whether or not segment j has been determined to be safe; segments are considered unsafe until they are scanned by a UAV. Given these variables, the desired properties to be verified include:

1. $\Diamond(vipLoc = 15)$
2. $\Box(vipLoc = uavLoc_1 \vee vipLoc = uavLoc_2 \vee vipLoc = uavLoc_3)$
3. $\Box((vipLoc = j) \rightarrow safeS_j)$ for $j = \{1,2,\ldots,15\}$

Property 1 states that the VIP should eventually reach the goal. Property 2 states that at least one of the UAVs should be at the same segment as the VIP at all times, or that the VIP should always be "protected." Property 3 states that segment at which the VIP is located should always be safe.

4.3.3.1 The Model

The model used to test mission plans for this scenario consists of four processes. The first is Planner (), which is responsible for constructing mission plans. Here, this involves determining the order in which the UAVs and VIP should visit certain road segments. The mission plans stored in Planner () could be generated by a human, an automation routine, or a combination of both.

The second process is Uav(), which models the behavior of the UAVs. It takes an input argument n that is used to assign a number to the UAV. Here, the process is instantiated once for each of the three UAVs using $n = \{0,1,2\}$. Each instance accesses an associated list of road segments generated by the Planner () and simulates scanning them. When a segment is scanned by a UAV, the UAV records whether or not the segment is safe. Segments are presumed to be unsafe until they are scanned by a UAV.

The third process is Vip(), which models the behavior of the VIP. Like the Uav() process, it takes a list of road segments generated by Planner () and simulates traveling over them.

The fourth process is Clock(), which is used to synchronize the other processes. Specifically, it implements a time step mechanism that is used to determine how long it takes for a UAV or the VIP to traverse a given road segment. Here, it is assumed that the UAVs and VIP move at the same constant speed. Thus, a fixed number of time steps is required to travel a particular road segment.

The model makes use of several global variables given in Listing 4.39. The first three variables correspond to $safeS_j$, $uavLoc_i$, and $vipLoc$. The next set of statements declare several **byte** queues of length 8 using the macros defined in Sect. 4.2.2.8. There is one queue for each UAV and one for the VIP. These are used by the Planner () to store ordered lists of road segments, which are then visited by the UAVs and VIP.

The last set of **bit** variables is used by the Clock() process, shown in Listing 4.40, to synchronize the actions of all other processes and establish a time step mechanism. Clock() begins by setting plannerReady to 1, enabling the Planner (). Once the Planner () is finished, it sets plannerReady to 0. The Clock() then sets the synchronization bits used by the other processes: uavReady[0], uavReady[1], uavReady[2], and vipReady. When these processes are finished, they set the corresponding **bit** variables to 0, and the Clock() is once again enabled. It then prints a status message and loops back to the beginning, repeating the process.

Listing 4.39 Global variables used in the UAV mission planning scenario

```
bit safeS[16] = 0;
byte uavLoc[3] = 0;
byte vipLoc = 0;

/* Declare queues of size 8, one for each UAV and the VIP */
DECLARE_QUEUE(byte,8);
QUEUE(byte,8) uavQueue[3];
QUEUE(byte,8) vipQueue;

/* Variables used for synchronization with the clock */
bit uavReady[3];
bit vipReady;
bit plannerReady;
```

Listing 4.40 The Clock() process

```
proctype Clock() {

 do
   ::
      /* Let the planner go first and wait for it to finish */
      plannerReady = 1;
      plannerReady == 0 ->
      atomic {
              /* All other processes can go */
              uavReady[0] = 1; uavReady[1] = 1;
              uavReady[2] = 1; vipReady = 1;
      }
      /* Wait for all processes, then print a status message */
      (!vipReady && !plannerReady &&
      !uavReady[0] && !uavReady[1] && !uavReady[2]) ->
      printf("vipLoc:%d,uavLoc:[%d,%d,%d]\n",vipLoc,
               uavLoc[0],uavLoc[1],uavLoc[2]);
 od;
}
```

The Uav() process – shown in Listing 4.41 – waits until it is enabled by Clock(). It then checks to see if its queue contains any new road segment assignments. If so, it changes its location to the next segment and records whether or not the segment is safe using the inline macro SegmentSafe(s, safe) given in Listing 4.52 of the Appendix. It then looks up the number of time steps needed to traverse the segment using the inline macro SegmentTime(s,t) given in Listing 4.53 of the Appendix. This number is stored in the variable timeToGo. If there are no new assignments stored in the queue, the UAV stays at its current location and sets timeToGo to 1. After checking for new assignments, Uav() waits the number of time steps indicated by timeToGo. Once the required number of time steps has passed, the process loops back to the beginning and repeats. The Vip() process, shown in Listing 4.42, is similar to the Uav() process, except that it does not check the safety status of segments.

Listing 4.41 The Uav() process

```
proctype Uav(byte myNum) {

  byte timeToGo = 0;

uavLoop:

  /* Wait until the Clock() signals a new time step */
  uavReady[myNum] ->

  /* If the queue is not empty, move to the next road segment.
     Otherwise, stay at the current segment for one time step. */
  if
    :: atomic{ IS_EMPTY(uavQueue[myNum]) -> timeToGo = 1;}
    :: atomic{ !IS_EMPTY(uavQueue[myNum]) ->
                 /* Update this UAV's segment position */
                 DEQUEUE(uavQueue[myNum], uavLoc[myNum]);
                 /* Update safety status of the segment */
                 segmentSafe(uavLoc[myNum], safeS[uavLoc[myNum]]);
                 /* Get time needed to travel this segment */
                 segmentTime(uavLoc[myNum], timeToGo);
    }
  fi;

  /* Wait the specified number of time increments */
  do
    :: atomic{(uavReady[myNum] && timeToGo > 1) -> timeToGo--;
               uavReady[myNum] = 0;}
    :: atomic{(uavReady[myNum] && timeToGo == 1) -> timeToGo--;
                                   uavReady[myNum] = 0; break;}
  od;

  goto uavLoop;

}
```

Listing 4.42 The Vip() process

```
proctype Vip() {

  byte timeToGo = 0;

vipLoop:

  /* Wait until I can take my turn */
  vipReady ->

  /* If the queue is not empty, get the next segment.
     Otherwise, stay at the current segment. */
  if
    :: atomic{ IS_EMPTY(vipQueue) -> timeToGo = 1;}
    :: atomic{ !IS_EMPTY(vipQueue) ->
           /* Update the VIP's segment position */
           DEQUEUE(vipQueue, vipLoc);
           /* Get the time needed to travel this segment */
           segmentTime(vipLoc,timeToGo);
       }
  fi;

  /* Wait the specified number of time increments */
  do
    :: atomic{(vipReady && timeToGo > 1) -> timeToGo--;
                           vipReady = 0;}
    :: atomic{(vipReady && timeToGo == 1) -> timeToGo--;
                           vipReady = 0; break;}
  od;

  goto vipLoop;

}
```

The Planner () process takes the form shown in Listing 4.43. It initializes the Vip() process, three instances of the Uav() process, and the Clock() process. The Planner () then specifies mission plans by assigning road segments to the UAVs and VIP using the appropriate queues. When Planner () is done executing mission plans or when it is in an idle state, it must continually set plannerReady to 0 in order to avoid deadlock in the model.

Listing 4.43 General form of the Planner() process

```
active proctype Planner() {

  /* Start the VIP process */
  run Vip();

  /* Start 3 UAV processes numbered 0 through 3 */
  run Uav(0); run Uav(1); run Uav(2);

  /* Start the clock */
  run Clock();

  /* Specify plans */
  <use queues to signal assignments to the UAVs and VIP>
  plannerReady = 0;

  ...

  <use queues to signal assignments to the UAVs and VIP>
  plannerReady = 0;

  /* Enter an idle status that keeps the clock running */
  do
    :: plannerReady == 1 -> plannerReady = 0;
  od;
}
```

4.3.3.2 Checking Mission Plans

The following examples will demonstrate how the Planner() process is configured in three different cases. In the first, a UAV is dispatched to check a single path that leads to node n_{11}. After a delay, the VIP and another UAV are directed to travel down the same path. In the second case, two UAVs are dispatched to check the safety of three different paths. The remaining UAV and VIP then follow one of the safe paths. The third case uses a modified version of the Vip() process to determine whether or not a solution exists for a given set of safe segment conditions.

Given the implementation details of the model, the desired properties discussed in Sect. 4.3.3 are written as in Listing 4.44. Note that the property requiring that there always be at least one UAV with the VIP is only checked at the end of each time step. Also, the properties $\Box((vipLoc = j) \rightarrow safeS_j)$ for $j = \{1, 2, \ldots, 15\}$ are split into two sets, due to a size limitation in SPIN's **ltl** parser.

Listing 4.44 The LTL properties to be verified in the UAV mission planning scenario

```
/* Property 1 */
ltl GoalReached        {<>(vipLoc == 15)}

/* Property 2 */
#define protected(x)  (vipLoc == uavLoc[x])
#define sync           (!plannerReady && !vipReady && !uavReady[0]
                        && !uavReady[1] && !uavReady[2])
ltl VipProtected {[](sync ->
                 (protected(0) || protected(1) || protected(2)))}

/* Property 3 */
#define safe(x)    ((vipLoc == x) -> safeS[x])
ltl SegmentSafe1 {safe(1) && safe(2) && safe(3) && safe(4) &&
                  safe(5) && safe(6) && safe(7) && safe(8)}
ltl SegmentSafe2 {safe(9)  && safe(10) && safe(11) && safe(12) &&
                  safe(13) && safe(14) && safe(15)}
```

Checking a Single Path

In this case, UAV 0 is directed by the Planner() to scan the dotted blue path shown in Figure 4.6. After waiting two additional time steps, the Planner() then directs the VIP and UAV 1 to traverse the same path. Here, the Planner() process takes the form shown in Listing 4.45. Initially, assume all segments are listed as safe in the SegmentSafe() macro. When SPIN is used to perform verification of the model under this condition, all properties are verified. The output generated by running SPIN in simulation mode is given in Listing 4.46. This output shows that UAV 0, UAV1, and the VIP traverse segments s_0, s_1, s_3, s_5, s_9, s_{14}, and s_{15}, with UAV 1 and the VIP lagging behind UAV 0 by two time steps. UAV 2 remains at segment s_0.

Assume now that one of the segments on the path, e.g. segment s_9, is listed as unsafe in the SegementSafe() macro. SPIN produces the same output in simulation mode. However, in verification mode, checking the property SegmentSafe2 results in an error, indicating that the path taken was not safe. Note that the other properties can still be verified.

Listing 4.45 The Planner() process used to check a single path

```
active proctype Planner() {

  /* Start 3 UAV processes numbered 0 through 3 */
  run Uav(0); run Uav(1); run Uav(2);

  /* Start the VIP process */
  run Vip();

  /* Start the clock */
  run Clock();

  /* Create initial assignment for UAV 0 - patrol blue path */
  atomic{
    ENQUEUE(uavQueue[0],0); ENQUEUE(uavQueue[0],1);
    ENQUEUE(uavQueue[0],3); ENQUEUE(uavQueue[0],5);
    ENQUEUE(uavQueue[0],9); ENQUEUE(uavQueue[0],14);
    ENQUEUE(uavQueue[0],15);
  }

  /* Wait 2 clock ticks */
  plannerReady -> plannerReady = 0;
  plannerReady -> plannerReady = 0;

  /* On the third tick, submit plans for UAV 1 and the VIP */
  atomic{
   plannerReady ->
   /* Create assignment for UAV 1 - fly the blue path */
   ENQUEUE(uavQueue[1],0); ENQUEUE(uavQueue[1],1);
   ENQUEUE(uavQueue[1],3); ENQUEUE(uavQueue[1],5);
   ENQUEUE(uavQueue[1],9); ENQUEUE(uavQueue[1],14);
   ENQUEUE(uavQueue[1],15);

  /* Create assignment for the VIP to travel the blue path */
  ENQUEUE(vipQueue,0); ENQUEUE(vipQueue,1);
  ENQUEUE(vipQueue,3); ENQUEUE(vipQueue,5);
  ENQUEUE(vipQueue,9); ENQUEUE(vipQueue,14);
  ENQUEUE(vipQueue,15);

  plannerReady = 0;
  }

  /* For the rest of the simulation, the planner simply loops */
  do
    :: plannerReady -> plannerReady = 0;
  od;

}
```

Listing 4.46 The output from Planner() when checking a single path

```
vipLoc: 0, uavLoc: [0, 0, 0]
vipLoc: 0, uavLoc: [0, 0, 0]
vipLoc: 0, uavLoc: [0, 0, 0]
vipLoc: 0, uavLoc: [1, 0, 0]
vipLoc: 0, uavLoc: [1, 0, 0]
vipLoc: 1, uavLoc: [3, 1, 0]
vipLoc: 1, uavLoc: [3, 1, 0]
vipLoc: 3, uavLoc: [5, 3, 0]
vipLoc: 3, uavLoc: [5, 3, 0]
vipLoc: 5, uavLoc: [9, 5, 0]
vipLoc: 5, uavLoc: [9, 5, 0]
vipLoc: 9, uavLoc: [9, 9, 0]
vipLoc: 9, uavLoc: [14, 9, 0]
vipLoc: 9, uavLoc: [15, 9, 0]
vipLoc: 14, uavLoc: [15, 14, 0]
vipLoc: 15, uavLoc: [15, 15, 0]
vipLoc: 15, uavLoc: [15, 15, 0]
...
```

Checking Multiple Paths

In this case, the Planner() directs two UAVs to scan the solid green, dotted blue, and dashed red paths shown in Figure 4.6. The remaining UAV and VIP take the first path that is found to be safe. The specific form of the Planner() used in this case is shown in Listing 4.47. It first directs UAV 0 to scan the green path and UAV 1 to simultaneously scan the blue path. UAV 1 is anticipated to finish first; thus, the Planner() waits until UAV 1 is finished with its scan. If the scan reveals that the blue path was safe, the Planner() directs UAV 2 and the VIP to travel this path. If it was not safe, Planner() directs UAV 1 to start scanning the red path. The Planner() then waits for UAV 0 to finish scanning the green path. If the green path was safe, it directs the VIP and UAV 2 to use it. Otherwise, it waits for UAV 1 to finish scanning the red path. If the red path was safe, it directs the VIP and UAV 2 to use it. Otherwise, the end of the Planner() process is reached, and there are no safe paths.

The path used by the VIP depends on which segments are safe. For instance, assume that segment s_9 is the only unsafe segment. The simulation output from SPIN in this case is shown in Listing 4.48. UAV 0 scans the green path, and UAV 1 scans the blue path. Because UAV 1 discovers that the blue path is unsafe, it begins scanning the red path. Once UAV 0 finishes scanning the green path and finds it to be safe, the VIP and UAV 2 use this path to reach segment s_{15}, and all properties are verified. Note that if certain combinations of segments are unsafe, e.g. s_{15} or s_8 and s_{14}, then the GoalReached property <>(vipLoc == 15) is violated. In cases such as this, no safe solution exists.

Listing 4.47 The Planner() process used when checking a multiple paths

```
active proctype Planner() {

  run Uav(0); run Uav(1); run Uav(2); run Vip(); run Clock();

  /*Direct UAVs 0 and 1 to scan green and blue paths. Wait for UAV 1 to finish*/
  ENQUEUE(uavQueue[0],0); ENQUEUE(uavQueue[0],1); ENQUEUE(uavQueue[0],4);
  ENQUEUE(uavQueue[0],8); ENQUEUE(uavQueue[0],15); ENQUEUE(uavQueue[1],0);
  ENQUEUE(uavQueue[1],1); ENQUEUE(uavQueue[1],3); ENQUEUE(uavQueue[1],5);
  ENQUEUE(uavQueue[1],9); ENQUEUE(uavQueue[1],14); ENQUEUE(uavQueue[1],15);

  do
    :: plannerReady && !IS_EMPTY(uavQueue[1]) -> plannerReady = 0;
    :: plannerReady && IS_EMPTY(uavQueue[1]) -> break;
  od;

  /* If safe, send the VIP and UAV 2 on the blue path and stop
     planning. Else, have UAV 1 scan the red path. Wait for UAV 0 to finish */
  if
    :: (safeS[0] && safeS[1] && safeS[3] && safeS[5] && safeS[9]
        && safeS[14] && safeS[15]) ->
       ENQUEUE(uavQueue[2],0); ENQUEUE(uavQueue[2],1); ENQUEUE(uavQueue[2],3);
       ENQUEUE(uavQueue[2],5); ENQUEUE(uavQueue[2],9); ENQUEUE(uavQueue[2],14);
       ENQUEUE(uavQueue[2],15); ENQUEUE(vipQueue,0); ENQUEUE(vipQueue,1);
       ENQUEUE(vipQueue,3); ENQUEUE(vipQueue,5); ENQUEUE(vipQueue,9);
       ENQUEUE(vipQueue,14); ENQUEUE(vipQueue,15);
       goto plannerLoop;
    :: else ->
       ENQUEUE(uavQueue[1],15); ENQUEUE(uavQueue[1],14);ENQUEUE(uavQueue[1],13);
       ENQUEUE(uavQueue[1],11); ENQUEUE(uavQueue[1],10);ENQUEUE(uavQueue[1],2);
  fi;

  do
    :: plannerReady && !IS_EMPTY(uavQueue[0]) -> plannerReady = 0;
    :: plannerReady && IS_EMPTY(uavQueue[0]) -> break;
  od;

  /* If safe, send the VIP and UAV 2 on the green path and stop planning.
     Else, wait until UAV 1 is done scanning the red path. */
  if
    :: (safeS[0] && safeS[1] && safeS[4] && safeS[8] && safeS[15]) ->
       ENQUEUE(uavQueue[2],0); ENQUEUE(uavQueue[2],1); ENQUEUE(uavQueue[2],4);
       ENQUEUE(uavQueue[2],8); ENQUEUE(uavQueue[2],15);
       ENQUEUE(vipQueue,0); ENQUEUE(vipQueue,1); ENQUEUE(vipQueue,4);
       ENQUEUE(vipQueue,8); ENQUEUE(vipQueue,15);
       goto plannerLoop;
    :: else -> skip;
  fi;

  do
    :: plannerReady && !IS_EMPTY(uavQueue[1]) -> plannerReady = 0;
    :: plannerReady && IS_EMPTY(uavQueue[1]) -> break;
  od;

  /* If safe, send the VIP and UAV 2 on the red path. Else, no paths are safe */
  if
    :: (safeS[0] && safeS[2] && safeS[10] && safeS[11] &&
        safeS[13] && safeS[14] && safeS[15]) ->
       ENQUEUE(uavQueue[2],0); ENQUEUE(uavQueue[2],2); ENQUEUE(uavQueue[2],10);
       ENQUEUE(uavQueue[2],11); ENQUEUE(uavQueue[2],13);
       ENQUEUE(uavQueue[2],14); ENQUEUE(uavQueue[2],15); ENQUEUE(vipQueue,0);
       ENQUEUE(vipQueue,2); ENQUEUE(vipQueue,10); ENQUEUE(vipQueue,11);
       ENQUEUE(vipQueue,13); ENQUEUE(vipQueue,14); ENQUEUE(vipQueue,15);
       goto plannerLoop;
    :: else -> printf("None of the paths are safe\n");
  fi;

plannerLoop:
  do
    :: plannerReady -> plannerReady = 0;
  od;
}
```

Listing 4.48 Output from the Planner() when checking a multiple paths

```
vipLoc: 0, uavLoc: [0, 0, 0]
vipLoc: 0, uavLoc: [0, 0, 0]
vipLoc: 0, uavLoc: [0, 0, 0]
vipLoc: 0, uavLoc: [1, 1, 0]
vipLoc: 0, uavLoc: [1, 1, 0]
vipLoc: 0, uavLoc: [4, 3, 0]
vipLoc: 0, uavLoc: [4, 3, 0]
vipLoc: 0, uavLoc: [4, 5, 0]
vipLoc: 0, uavLoc: [8, 5, 0]
vipLoc: 0, uavLoc: [8, 9, 0]
vipLoc: 0, uavLoc: [8, 9, 0]
vipLoc: 0, uavLoc: [8, 9, 0]
vipLoc: 0, uavLoc: [8, 14, 0]
vipLoc: 0, uavLoc: [8, 15, 0]
vipLoc: 0, uavLoc: [8, 15, 0]
vipLoc: 0, uavLoc: [8, 15, 0]
vipLoc: 0, uavLoc: [15, 15, 0]
vipLoc: 0, uavLoc: [15, 15, 0]
vipLoc: 0, uavLoc: [15, 15, 0]
vipLoc: 0, uavLoc: [15, 15, 0]
vipLoc: 1, uavLoc: [15, 15, 1]
vipLoc: 1, uavLoc: [15, 14, 1]
vipLoc: 4, uavLoc: [15, 13, 4]
vipLoc: 4, uavLoc: [15, 13, 4]
vipLoc: 4, uavLoc: [15, 13, 4]
vipLoc: 8, uavLoc: [15, 13, 8]
vipLoc: 8, uavLoc: [15, 11, 8]
vipLoc: 8, uavLoc: [15, 11, 8]
vipLoc: 8, uavLoc: [15, 10, 8]
vipLoc: 8, uavLoc: [15, 10, 8]
vipLoc: 8, uavLoc: [15, 10, 8]
vipLoc: 8, uavLoc: [15, 2, 8]
vipLoc: 8, uavLoc: [15, 2, 8]
vipLoc: 15, uavLoc: [15, 2, 15]
vipLoc: 15, uavLoc: [15, 2, 15]
...
```

Checking That a Solution Exists

With some modifications to the model, the Planner() and Vip() processes can be used to check whether or not a solution exists given a fixed set of safety conditions for the road segments. First, a new global variable vipNode is added to model. This variable tracks the node at which the VIP is located after it reaches the end of a segment. Then, the Planner() takes the form shown in Listing 4.49. It directs the VIP to choose a random segment to travel based on the road network's topology and the VIP's current node location. The modified Vip() process, shown in Listing 4.50, gets new assignments from the Planner(). When the VIP moves to a new segment, it checks to see whether or not the segment is safe. If it is not, it stays at its current node location. If it is, it updates its node location. Then, the process repeats. Note that the Clock() process is modified to only use the plannerReady and vipReady variables; uavReady[0], uavReady[1], and uavReady[2] have been removed.

Listing 4.49 The Planner() process used to determine whether a solution exists

```
active proctype Planner() {

  /* Start the VIP and clock */
  run Vip(); run Clock();

  /* Allow the VIP to choose a random assignment at each node. */
chooseNextSegment:
  if
   :: plannerReady && vipNode == 0 -> ENQUEUE(vipQueue,0);
   :: plannerReady && vipNode == 1 -> ENQUEUE(vipQueue,1);
   :: plannerReady && vipNode == 1 -> ENQUEUE(vipQueue,2);
   :: plannerReady && vipNode == 2 -> ENQUEUE(vipQueue,2);
   :: plannerReady && vipNode == 2 -> ENQUEUE(vipQueue,7);
   :: plannerReady && vipNode == 2 -> ENQUEUE(vipQueue,10);
   :: plannerReady && vipNode == 3 -> ENQUEUE(vipQueue,1);
   :: plannerReady && vipNode == 3 -> ENQUEUE(vipQueue,3);
   :: plannerReady && vipNode == 3 -> ENQUEUE(vipQueue,4);
   :: plannerReady && vipNode == 4 -> ENQUEUE(vipQueue,3);
   :: plannerReady && vipNode == 4 -> ENQUEUE(vipQueue,5);
   :: plannerReady && vipNode == 4 -> ENQUEUE(vipQueue,6);
   :: plannerReady && vipNode == 5 -> ENQUEUE(vipQueue,4);
   :: plannerReady && vipNode == 5 -> ENQUEUE(vipQueue,6);
   :: plannerReady && vipNode == 5 -> ENQUEUE(vipQueue,8);
   :: plannerReady && vipNode == 6 -> ENQUEUE(vipQueue,5);
   :: plannerReady && vipNode == 6 -> ENQUEUE(vipQueue,7);
   :: plannerReady && vipNode == 6 -> ENQUEUE(vipQueue,9);
   :: plannerReady && vipNode == 7 -> ENQUEUE(vipQueue,10);
   :: plannerReady && vipNode == 7 -> ENQUEUE(vipQueue,11);
   :: plannerReady && vipNode == 7 -> ENQUEUE(vipQueue,12);
   :: plannerReady && vipNode == 8 -> ENQUEUE(vipQueue,11);
   :: plannerReady && vipNode == 8 -> ENQUEUE(vipQueue,12);
   :: plannerReady && vipNode == 8 -> ENQUEUE(vipQueue,13);
   :: plannerReady && vipNode == 9 -> ENQUEUE(vipQueue,9);
   :: plannerReady && vipNode == 9 -> ENQUEUE(vipQueue,14);
   :: plannerReady && vipNode == 10-> ENQUEUE(vipQueue,15);
  fi;
  plannerReady = 0;
  goto chooseNextSegment;

}
```

In this case, the negation of the property GoalReached, i.e. $\Box\neg(vipLoc == 15)$, can be checked. If the property is verified, then no solution exists. For instance, if s_0 is unsafe or s_8, s_9, and s_{13} are unsafe, then the property is verified and there are no safe paths to s_{15}. If verification returns an error, then the error trail reveals a path that allows the VIP to safely reach segment s_{15}. For instance, if segments s_3, s_6, and s_9 are unsafe, then SPIN provides the following path as a counterexample:

$s_0 \rightarrow s_1 \rightarrow s_1 \rightarrow s_2 \rightarrow s_7 \rightarrow s_5 \rightarrow s_5 \rightarrow s_7 \rightarrow s_{10} \rightarrow s_{11} \rightarrow s_{12} \rightarrow s_{13} \rightarrow s_{14} \rightarrow s_{15}$. Note that this is not an optimal path; the solution includes some backtracking. However, it could be refined to produce a more reasonable solution.

Listing 4.50 The modified Vip() process used to determine whether a solution exists

```
proctype Vip() {

  bit isSafe;

vipLoop:

  /* Wait until I can take my turn to get the next segment */
  vipReady && !IS_EMPTY(vipQueue) -> DEQUEUE(vipQueue, vipLoc);

  /* Check that the segment is safe. If not, stay at the
     current location for the rest of the simulation */
  segmentSafe(vipLoc, isSafe);
  do
    :: isSafe -> break;
    :: !isSafe -> vipReady = 0; goto vipLoop;
  od;

  if
    :: vipNode == 0  && vipLoc == 0  -> vipNode = 1;
    :: vipNode == 1  && vipLoc == 1  -> vipNode = 3;
    :: vipNode == 1  && vipLoc == 2  -> vipNode = 2;
    :: vipNode == 2  && vipLoc == 2  -> vipNode = 1;
    :: vipNode == 2  && vipLoc == 7  -> vipNode = 6;
    :: vipNode == 2  && vipLoc == 10 -> vipNode = 7;
    :: vipNode == 3  && vipLoc == 1  -> vipNode = 1;
    :: vipNode == 3  && vipLoc == 3  -> vipNode = 4;
    :: vipNode == 3  && vipLoc == 4  -> vipNode = 5;
    :: vipNode == 4  && vipLoc == 3  -> vipNode = 3;
    :: vipNode == 4  && vipLoc == 5  -> vipNode = 6;
    :: vipNode == 4  && vipLoc == 6  -> vipNode = 5;
    :: vipNode == 5  && vipLoc == 4  -> vipNode = 3;
    :: vipNode == 5  && vipLoc == 6  -> vipNode = 4;
    :: vipNode == 5  && vipLoc == 8  -> vipNode = 10;
    :: vipNode == 6  && vipLoc == 5  -> vipNode = 4;
    :: vipNode == 6  && vipLoc == 7  -> vipNode = 2;
    :: vipNode == 6  && vipLoc == 9  -> vipNode = 9;
    :: vipNode == 7  && vipLoc == 10 -> vipNode = 2;
    :: vipNode == 7  && vipLoc == 11 -> vipNode = 8;
    :: vipNode == 7  && vipLoc == 12 -> vipNode = 8;
    :: vipNode == 8  && vipLoc == 11 -> vipNode = 7;
    :: vipNode == 8  && vipLoc == 12 -> vipNode = 7;
    :: vipNode == 8  && vipLoc == 13 -> vipNode = 9;
    :: vipNode == 9  && vipLoc == 9  -> vipNode = 6;
    :: vipNode == 9  && vipLoc == 14 -> vipNode = 10;
    :: vipNode == 10 && vipLoc == 15 -> vipNode = 11;
  fi;
  vipReady = 0;

  goto vipLoop;
}
```

4.4 Discussion

The preceding applications demonstrate how a model checker such as SPIN can be used in the design and verification of UAV related systems. They also demonstrate some of the strengths and weaknesses of modeling languages such as PROMELA. On one hand, PROMELA provides a natural way to model systems comprised of multiple assets, i.e. by defining multiple processes. The same process can be instantiated multiple times to represent multiple instances of a homogeneous asset, as in the UAV cooperative controller of Sect. 4.3.1. Different processes can be instantiated to represent heterogeneous assets, such as the UAVs and VIP in Sect. 4.3.3. On the other hand, one must take care that multiple processes execute in a realistic fashion. For instance, the UAVs in Sect. 4.3.1 should execute asynchronously but fairly, i.e. no process should "starve." To address this issue in SPIN, one can run the verifier using weak fairness conditions. However, one must take care that the processes do not contain blocking statements; otherwise, weak fairness has no effect. Another option is to use synchronization events, as in Sect. 4.3.2 or Sect. 4.3.3. However, one must avoid modeling errors that lead to deadlock when using such mechanisms.

Modeling time in SPIN is possible though sometimes difficult. Establishing .5 second "weak epochs" in Sect. 4.3.2 requires careful coordination of several variables. Had the sensors used epochs of different lengths, the model would have been much more complicated. Similarly, in the mission planning scenario of Sect. 4.3.3, the UAVs and VIP were assumed to travel at the same speed. Thus, a single macro providing a simple lookup table was used to determine how many time steps were required to traverse a road segment. Had this assumption not been made, the model would have required a more complex lookup table.

A well-known weakness of model checking comes from the state space explosion problem. The number of execution paths to be checked in a model grows exponentially with the number of possible variables and process execution interleavings. This can result in tight limitations on model sizes. For instance, the model in Sect. 4.3.1 includes only two UAVs and four ground sensors, and the model in Sect. 4.3.2 includes only three ground sensors. When more UAVs or ground sensors are added to these models, the verification times increase substantially. Models of arbitrary size are generally intractable.

Finally, the models presented here were designed at a relatively high level of abstraction. For instance, a real implementation of the UAV cooperative controller in Sect. 4.3.1 would utilize optimization routines to compute more precise orderings for ground sensor patrol routes. As long as these optimization routines result in executions that are a subset of those of exhibited by the verified model, the optimized system will still have the desired properties. However, one must take care that optimized implementation is faithful to the model.

4.5 Current and Future Directions

While formal methods such as model checking were originally developed to address simple software and hardware design and verification problems, the number and scope of application problems is growing. The following discusses some new areas of research that incorporate model checking and other formal methods. These areas include hybrid and embedded systems, model-driven design, statistical model checking, abstraction techniques, human-automation interfaces, and biological model checking.

Hybrid systems are those that include both discrete and continuous dynamics [1]. This includes embedded systems, i.e. physical systems controlled by discrete logic. For instance, consider an engine whose fuel injection system is regulated by a microprocessor. The continuous dynamics describing the amount of fuel being injected into the engine can be modified by discrete decisions made by the microprocessor. The system is thus described by a state consisting of both discrete *control modes* and continuous variables [14]. The discrete control modes can be modeled by a graph, with vertices representing the modes and edges representing transitions or *jumps* between modes. Changes in the continuous variables are generally modeled by differential equations. The discrete control mode determines the form of the differential equations governing the continuous variables, and transitions between discrete control modes are governed by conditions that depend on the continuous variables.

General issues in hybrid systems include simulation, determining existence and uniqueness of solutions, ensuring non-Zeno behavior, determining whether certain states are reachable from other states, analyzing the composition of multiple hybrid systems, and verifying that a hybrid system has certain desired behaviors. The last involves the use of model checking [15, 44]. However, since non-trivial hybrid systems have an infinite number of states, the states must be divided into a finite number of regions, each of which can be characterized by a set of properties of interest. Then, desired system properties can be specified, e.g. using timed computation tree logic [1]. Hybrid systems and model checking have applications in areas such as collision avoidance. For instance, Ref. [30] discusses design and verification of a hybrid cooperative controller for platoons of vehicles traveling on a highway, and Ref. [42] discusses design and verification of a hybrid controller for an evader in an adversarial pursuer-evader scenario.

A related area is model-driven design of large scale systems, including hybrid and embedded systems. Design and verification of such systems is difficult because they contain numerous complex and heterogeneous components. Model-driven design techniques that focus on the use of formal methods during the design phase often result in more robust designs and a more thorough and organized verification process [10]. However, applying formal methods to such systems can be difficult. Formal methods tend to be tailored to smaller, homogenous systems. Thus, research in this area includes studying how systems can be decomposed or broken into hierarchies

[20, 36]; how to develop open and inter-operable specifications, architectures, and design languages at the right level of abstraction [25]; how to combine different formal methods such as model checking and automated proof checking [32]; and how to manage and cross-check verification results obtained from different formal methods as part of a *heterogenous verification* process [26].

More specific techniques are also being developed to support model checking for complex systems, especially those that would be difficult or cumbersome to describe mathematically. For such systems, one can consider abstraction techniques, in which an approximation of the system that retains only a small number of relevant variables is checked [29]. Methods that aid in the automatic generation of such abstractions are currently being researched [7, 8]. Another technique that is being promoted for verification of complex systems is statistical model checking, in which a sampling of system simulations is used to determine the probability that a system meets a desired specification, within certain confidence bounds [27, 37, 46].

Model checking has also been applied to the problem of human-automation interface design and verification. Ref. [6] provides a overview of early work in this area. For human interfaces, there are at least three classes properties that should be verified. These include *visibility*, which concerns what is shown on the user interface and how it is perceived by the user; *reachability*, which concerns what can be done with the interface and how the user does it; and general behaviors of the interface, including properties related to the current state. Visibility properties to be verified might include checking that certain events have appropriate feedback; reachability properties to be verified might include checking that actions can be undone; and general behaviors to be verified might include checking that the same event always has the same effect in a given mode or that some property of the interface holds in every state. A related topic is checking for "automation surprises," or states in the automation that are significantly different than what the user expects [34]. Such states can be identified by comparing a formal model of the user's view of the system to a formal model of the system itself and checking for discrepancies.

Finally, model checking has recently been applied to problems in biology. Certain biological systems can be represented by formal computational models. When such models are available, model checking can be used in hypothesis testing [11]. Suppose several different biological mechanisms are proposed to explain a set of observations. A computational model for each mechanism can be developed, and model checking can be applied to see which models agree with the observations and which do not. This narrows the field of valid hypotheses. Model checking is also important for verification of medical devices, to ensure that they have the expected behavior and that they safely interact with the patient and with other devices [28]. Software development practices that lead to safe and verifiable devices are of critical importance [23]. Also, many medical devices are complex, embedded systems; thus, new techniques such as statistical model checking are being applied to problems in this area [22].

Appendix

Listing 4.51 The AssignUGSs() inline macro used in the UAV cooperative control scenario

```
inline AssignUGSs(uavActive, ugsActive, ugsList0, ugsList1) {
 atomic {
  /*Initialize UGS statuses and assignment lists to 0*/
  SET_ALL_0(ugsActive); SET_ALL_0(ugsList0); SET_ALL_0(ugsList1);

  if /*Set a UGS to be active. Assign to UAV 0 if it's active*/
    :: IS_1(uavActive,0) -> SET_1(ugsActive,0); SET_1(ugsList0,0);
    :: IS_1(uavActive,0) -> SET_1(ugsActive,1); SET_1(ugsList0,1);
    :: IS_1(uavActive,0) -> SET_1(ugsActive,2); SET_1(ugsList0,2);
    :: IS_1(uavActive,0) -> SET_1(ugsActive,3); SET_1(ugsList0,3);
    :: else -> skip;
  fi;

  if /*Set another UGS to be active. Assign to UAV 1 if active*/
    ::IS_1(uavActive,1) && IS_0(ugsActive,0)->SET_1(ugsActive,0);
                                              SET_1(ugsUav1,0);
    ::IS_1(uavActive,1) && IS_0(ugsActive,1)->SET_1(ugsActive,1);
                                              SET_1(ugsUav1,1);
    ::IS_1(uavActive,1) && IS_0(ugsActive,2)->SET_1(ugsActive,2);
                                              SET_1(ugsUav1,2);
    ::IS_1(uavActive,1) && IS_0(ugsActive,3)->SET_1(ugsActive,3);
                                              SET_1(ugsUav1,3);
    ::else -> skip;
  fi;

   do /* Randomly set some or no other UGSs to be active */
    :: IS_1(uavActive,0) && IS_0(ugsActive,0)->SET_1(ugsList0,0);
                                               SET_1(ugsActive,0);
    :: IS_1(uavActive,1) && IS_0(ugsActive,0)->SET_1(ugsList1,0);
                                               SET_1(ugsActive,0);
    :: IS_1(uavActive,0) && IS_0(ugsActive,1)->SET_1(ugsList0,1);
                                               SET_1(ugsActive,1);
    :: IS_1(uavActive,1) && IS_0(ugsActive,1)->SET_1(ugsList1,1);
                                               SET_1(ugsActive,1);
    :: IS_1(uavActive,0) && IS_0(ugsActive,2)->SET_1(ugsList0,2);
                                               SET_1(ugsActive,2);
    :: IS_1(uavActive,1) && IS_0(ugsActive,2)->SET_1(ugsList1,2);
                                               SET_1(ugsActive,2);
    :: IS_1(uavActive,0) && IS_0(ugsActive,3)->SET_1(ugsList0,3);
                                               SET_1(ugsActive,3);
    :: IS_1(uavActive,1) && IS_0(ugsActive,3)->SET_1(ugsList1,3);
                                               SET_1(ugsActive,3);
    :: skip -> break;
  od;
 }
}
```

Listing 4.52 The SegmentSafe() inline macro used to check whether or not a given segment is safe in the UAV mission planning scenario

```
inline SegmentSafe(s,safe) {
  if
    :: s == 0  -> safe = 1;
    :: s == 1  -> safe = 1;
    :: s == 2  -> safe = 1;
    :: s == 3  -> safe = 1;
    :: s == 4  -> safe = 1;
    :: s == 5  -> safe = 1;
    :: s == 6  -> safe = 1;
    :: s == 7  -> safe = 1;
    :: s == 8  -> safe = 0;
    :: s == 9  -> safe = 0;
    :: s == 10 -> safe = 1;
    :: s == 11 -> safe = 1;
    :: s == 12 -> safe = 1;
    :: s == 13 -> safe = 1;
    :: s == 14 -> safe = 1;
    :: s == 15 -> safe = 1;
  fi;
}
```

Listing 4.53 The SegmentTime() inline macro used to look up the number of time steps required to traverse road segments in the UAV mission planning scenario

```
inline segmentTime(s,t) {
  if
    :: s == 0  -> t = 3;
    :: s == 1  -> t = 2;
    :: s == 2  -> t = 4;
    :: s == 3  -> t = 2;
    :: s == 4  -> t = 3;
    :: s == 5  -> t = 2;
    :: s == 6  -> t = 5;
    :: s == 7  -> t = 5;
    :: s == 8  -> t = 8;
    :: s == 9  -> t = 3;
    :: s == 10 -> t = 3;
    :: s == 11 -> t = 2;
    :: s == 12 -> t = 3;
    :: s == 13 -> t = 4;
    :: s == 14 -> t = 1;
    :: s == 15 -> t = 4;
  fi;
}
```

References

1. Alur, R., Courcoubetis, C., Halbwachs, N., Henzinger, T.A., Ho, P.H., Nicollin, X., Olivero, A., Sifakis, J., Yovine, S.: The algorithmic analysis of hybrid systems. Theoretical Computer Science 138(1), 3–34 (1995)
2. Baier, C., Joost-Pieter, K.: Principles of Model Checking. The MIT Press (2008)
3. Boehm, B.: Software and Its Impact: A Quantitative Assessment. Rand Corp. (1972)
4. Boehm, B.: Software engineering economics. IEEE Transactions on Software Engineering SE-10(1), 4–21 (1984)
5. Boehm, B.: A view of 20th and 21st century software engineering. In: Proceedings of the 28th International Conference on Software Engineering, pp. 12–29. ACM (2006)
6. Campos, J.C., Harrison, M.D.: Formally verifying interactive systems: A review. In: Proceedings of the Fourth International Eurographics Workshop on Design, Specification, and Verification of Interactive Systems, pp. 109–124 (1997)
7. Chauhan, P., Clarke, E., Kukula, J., Sapra, S., Veith, H., Wang, D.: Automated abstraction refinement for model checking large state spaces using SAT based conflict analysis. In: Aagaard, M.D., O'Leary, J.W. (eds.) FMCAD 2002. LNCS, vol. 2517, pp. 33–51. Springer, Heidelberg (2002)
8. Clarke, E., Grumberg, O., Jha, S., Lu, Y., Veith, H.: Counterexample-guided abstraction refinement for symbolic model checking. Journal of the ACM (JACM) 50(5), 752–794 (2003)
9. Dahm, W.: Report on technology horizons: A vision for Air Force science & technology during 2010-2030. Technical report, United States Air Force (2010)
10. Edwards, S., Lavagno, L., Lee, E.A., Sangiovanni-Vincentelli, A.: Design of embedded systems: Formal models, validation, and synthesis. Proceedings of the IEEE 85(3), 366–390 (1997)
11. Fisher, J., Henzinger, T.A.: Executable cell biology. Nature Biotechnology 25(11), 1239–1249 (2007)
12. Fisher, M., Bordini, R.H., Hirsch, B., Torroni, P.: Computational logics and agents: A roadmap of current technologies and future trends. Computational Intelligence 23(1), 61–91 (2007)
13. Gerth, R.: Concise promela reference (1997), `http://spinroot.com/spin/Man/Quick.html` (accessed April 2011)
14. Henzinger, T.A.: The theory of hybrid automata. In: Proceedings of the Eleventh Annual IEEE Symposium on Logic in Computer Science (LICS 1996), pp. 278–292. IEEE (1996)
15. Henzinger, T.A., Ho, P.H., Wong-Toi, H.: HyTech: A model checker for hybrid systems. International Journal on Software Tools for Technology Transfer (STTT) 1(1), 110–122 (1997)
16. Hoare, C.A.R.: An axiomatic basis for computer programming. Communications of the ACM 12(10), 576–580 (1969)
17. Holzmann, G.J.: The model checker SPIN. IEEE Transactions on Software Engineering 23(5), 279–294 (1997)
18. Holzmann, G.J.: The SPIN Model Checker: Primer and Reference Manual. Addison Wesley Publishing Company (2004)
19. Holzmann, G.J., Peled, D.: An improvement in formal verification. In: Proceedings of the FORTE 1994 Conference (1994)
20. Horowitz, R., Varaiya, P.: Control design of an automated highway system. Proceedings of the IEEE 88(7), 913–925 (2000)

21. Iglesias, C.A., Garijo, M., González, J.C.: A survey of agent-oriented methodologies. In: Papadimitriou, C., Singh, M.P., Müller, J.P. (eds.) ATAL 1998. LNCS (LNAI), vol. 1555, pp. 317–330. Springer, Heidelberg (1999)
22. Jha, S.K., Clarke, E.M., Langmead, C.J., Legay, A., Platzer, A., Zuliani, P.: A bayesian approach to model checking biological systems. In: Degano, P., Gorrieri, R. (eds.) CMSB 2009. LNCS (LNBI), vol. 5688, pp. 218–234. Springer, Heidelberg (2009)
23. Jiang, Z., Pajic, M., Connolly, A., Dixit, S., Mangharam, R.: Real-time heart model for implantable cardiac device validation and verification. In: 22nd Euromicro Conference on Real-Time Systems (ECRTS 2010), pp. 239–248. IEEE (2010)
24. Karaman, S., Frazzoli, E.: Vehicle routing with linear temporal logic specifications: Applications to multi-UAV mission planning. In: Proceedings of the AIAA Conference on Guidance, Navigation, and Control (2008)
25. Kent, S.: Model driven engineering. In: Butler, M., Petre, L., Sere, K. (eds.) IFM 2002. LNCS, vol. 2335, pp. 286–298. Springer, Heidelberg (2002)
26. Kumar, R., Krogh, B.H.: Heterogeneous verification of embedded control cystems. In: Proceedings of the IEEE American Control Conference, ACC 2006 (2006)
27. Kwiatkowska, M., Norman, G., Parker, D.: PRISM 2.0: A tool for probabilistic model checking. In: Proceedings of the First International Conferece on Quantitative Evaluation of Systems (QEST 2004), pp. 322–323. IEEE (2004)
28. Lee, I., Pappas, G.J., Cleaveland, R., Hatcliff, J., Krogh, B.H., Lee, P., Rubin, H., Sha, L.: High-confidence medical device software and systems. IEEE Computer 39(4), 33–38 (2006)
29. Long, D.E.: Model Checking, Abstraction, and Compositional Verification. PhD thesis, Carnegie Mellon University (1993)
30. Lygeros, J., Godbole, D.N., Sastry, S.: Verified hybrid controllers for automated vehicles. IEEE Transactions on Automatic Control 43(4), 522–539 (1998)
31. Pneuli, A.: The temporal logic of programs. In: Proceedings of the 18th IEEE Symposium on Foundations of Computer Science (FOCS 1977), pp. 46–67. IEEE Computer Society Press (1977)
32. Rajan, S., Shankar, N., Srivas, M.K.: An integration of model checking with automated proof checking. In: Wolper, P. (ed.) CAV 1995. LNCS, vol. 939, pp. 84–97. Springer, Heidelberg (1995)
33. Royce, W.W.: Managing the development of large software systems. In: Proceedings of IEEE WESCON (1970)
34. Rushby, J.: Using model checking to help discover mode confusions and other automation surprises. Reliability Engineering and System Safety 75(2), 167–177 (2002)
35. Ruys, T.C.: Towards Effective Model Checking. PhD thesis, University of Twente (2001)
36. Schmidt, D.C.: Guest editor's introduction: Model-driven engineering. IEEE Computer 39(2), 25–31 (2006)
37. Sen, K., Viswanathan, M., Agha, G.: Statistical model checking of black-box probabilistic systems. In: Alur, R., Peled, D.A. (eds.) CAV 2004. LNCS, vol. 3114, pp. 202–215. Springer, Heidelberg (2004)
38. Sharp, C., Schaffert, S., Woo, A., Sastry, N., Karlof, C., Sastry, S., Culler, D.: Design and implementation of a sensor network system for vehicle tracking and autonomous interception. In: Proceedings of the 2nd European Workshop on Wireless Sensor Networks, pp. 93–107 (2005)
39. Sirigineedi, G., Tsourdos, A., White, B., Zbikowski, R.: Kripke modelling and model checking of a multiple UAV system monitoring road network. In: Proceedings of the AIAA Guidance, Navigation, and Control Conference (2010)

40. Tassey, G.: The economic impacts of inadequate infrastructure for software testing. Technical Report RTI Project Number 7007.011, National Institute of Standards and Technology (2002)
41. The Standish Group. CHAOS Chronicles Online (2007)
42. Tomlin, C.J., Mitchell, I., Bayen, A.M., Oishi, M.: Computational techniques for the verification of hybrid systems. Proceedings of the IEEE 91(7), 986–1001 (2003)
43. Vardi, M.Y.: Branching vs. Linear time: Final showdown. In: Margaria, T., Yi, W. (eds.) TACAS 2001. LNCS, vol. 2031, pp. 1–22. Springer, Heidelberg (2001)
44. Wongpiromsarn, T., Topcu, U., Ozay, N., Xu, H., Murray, R.M.: TuLiP: A software toolbox for receding horizon temporal logic planning. In: Proceedings of the 14th International Conference on Hybrid Systems: Computation and Control, pp. 313–314 (2011)
45. Wooldridge, M.J., Ciancarini, P.: Agent-oriented software engineering: The state of the art. In: Ciancarini, P., Wooldridge, M.J. (eds.) AOSE 2000. LNCS, vol. 1957, pp. 1–28. Springer, Heidelberg (2001)
46. Younes, H.L.S., Kwiatkowska, M., Norman, G., Parker, D.: Numerical vs. statistical probabilistic model checking. International Journal on Software Tools for Technology Transfer (STTT) 8(3), 216–228 (2006)

Chapter 5
Approximate Dynamic Programming Applied to UAV Perimeter Patrol

K. Krishnamoorthy, M. Park, S. Darbha, M. Pachter, and P. Chandler

Abstract. One encounters the *curse of dimensionality* in the application of dynamic programming to determine optimal policies for large scale controlled Markov chains. In this chapter, we consider a base perimeter patrol stochastic control problem. To determine the optimal control policy, one has to solve a Markov decision problem, whose large size renders exact dynamic programming methods intractable. So, we propose a state aggregation based approximate linear programming method to construct provably good sub-optimal policies instead. The state-space is partitioned and the optimal cost-to-go or value function is approximated by a constant over each partition. By minimizing a non-negative cost function defined on the partitions, one can construct an approximate value function which also happens to be an upper bound for the optimal value function of the original Markov chain. As a general result, we show that this approximate value function is *independent* of the non-negative cost function (or state dependent weights; as it is referred to in the literature) and moreover, this is the least upper bound that one can obtain, given the partitions. Furthermore, we show that the restricted system of linear inequalities also embeds a family of Markov chains of lower dimension, one of which can be used to construct a tight lower bound on the optimal value function. In general, the construction of the lower bound requires the solution to a combinatorial problem. But

K. Krishnamoorthy
Infoscitex Corporation, Dayton, OH 45431
e-mail: krishnak@ucla.edu

M. Park
Graduate Student, Texas A& M University, College Station, TX 77843

S. Darbha
Texas A& M University, College Station, TX 77843

M. Pachter
Air Force Institute of Technology, Wright-Patterson A.F.B., OH 45433

P. Chandler
Air Force Research Laboratory, Wright-Patterson A.F.B., OH 45433

F. Fahroo et al. (Eds.): *Recent Adv. in Res. on Unmanned Aerial Veh.*, LNCIS 444, pp. 119–146.
DOI: 10.1007/978-3-642-37694-8_5

the perimeter patrol problem exhibits a special structure that enables tractable linear programming formulations for both the upper and lower bounds. We demonstrate this and also provide numerical results that corroborate the efficacy of the proposed methodology.

5.1 Base Perimeter Patrol

We address the following base perimeter patrol problem: an Unmanned Aerial Vehicle (UAV) and a remotely located operator cooperatively perform the task of perimeter patrol. Alert stations consisting of Unattended Ground Sensors (UGSs) are located at key locations along the perimeter. Upon detection of an incursion in its sector, an alert is flagged by the UGS. The statistics of the alerts' arrival process is known and, as in queueing theory, is a Poisson process. A camera equipped UAV is on continuous patrol along the perimeter and is tasked with inspecting UGSs with alerts. On arrival to a alert flagged UGS, the UAV then orbits the UGS/alert station and a video feed is transmitted to the operator. The latter can steer the gimballed camera while looking for the cause of the alarm. Naturally, the longer a UAV dwells (loiters) at an alert site, the more information it gathers and transmits to the operator. The objective here is to maximize the information gained, and, at the same time, reduce the expected response time to an alert. The problem is simplified by considering discrete-time evolution equations for updating the position of the UAV and also by considering only a finite (fixed) number, m, of UGS locations. It is assumed that the UAV has access to real-time information about the status of alerts (whether they have been attended to or not) at each alert station.

The perimeter patrol problem as envisaged here, falls in the domain of discrete-time controlled queueing systems [36]. In general, a *queueing system* includes arriving customers, servers, and waiting lines/buffers, or, queues, for the customers awaiting service [19]. In the context of perimeter patrol, the "customers" are the flagged UGSs/alerts waiting to be serviced and the UAV is the server. In queueing theory, the queues/buffers could be of finite or infinite capacity. Here only unit/single buffer queueing is considered, for the UGS either flags an alert or it does not. Once it flags an alert, its state does not change, even if additional triggering events were to occur, until the flag is reset by a loitering UAV. Hence there is at most only one alert waiting at an alert site. So, the perimeter patrol problem as envisaged here, constitutes a multi-queue, single-server, unit-buffer queueing system with deterministic (inter station) travel and service times. Since the UAV is constantly on patrol or is servicing a triggered UGS, the framework considered here is analogous to the *cyclic polling system* shown in Fig. 5.1. The basic model of a cyclic polling system consists of separate queues with independent Poisson arrivals (say, at rate α) served by a single server in cyclic order. A thorough analysis of such systems, and the various applications thereof, can be found in [39]. In the queueing literature, the performance measures of interest have been primarily the average queue length and average customer waiting time for service. Significant effort has been expended in evaluating these two metrics for polling systems under varying assumptions on the service time statistics, buffer capacity and service discipline [37, 38]. One of the earliest applications of a polling model with a single buffer and deterministic travel and service times arose in the British cotton industry involving a patrolling machine repairman [26]. A related problem is the dynamic traveling repairman problem, where the stations are not restricted to being on a line segment or a closed path [5].

Traditionally, in queueing theory, including polling systems, the server's action, i.e., which station to move towards next, is considered to be the control variable [25]. But the service time itself is either considered to be a random variable with a known p.d.f., for e.g., an exponential distribution, or a deterministic constant. So for a single buffer polling system, an optimization problem can be set up with the objective of say, minimizing the service delay, or maximizing the message throughput [15, 6, 1, 20]. However, from a patrolling perspective, one is interested in the optimal service time in addition to the dynamic scheduling of the server's movement. The basic question then would be to decide how long the server/UAV should dwell at a triggered alert station/UGS, and also in which station to move towards next, upon completion of a service. So, in this work, the primary interest is in determining the optimal service time i.e., the dwell time spent by the UAV at an alert site. In addition, the optimal controller also determines which station the UAV should head toward next, upon completion of an alert service. The present work considers a metric which trades off the information gained by the UAV as a function of the time spent loitering at the alert sites gathering information on the nature of the alerts, and the time taken by the UAV to respond to other alerts, that is, the time it takes the UAV to reach a triggered UGS. Thus, the metric increases monotonically with the

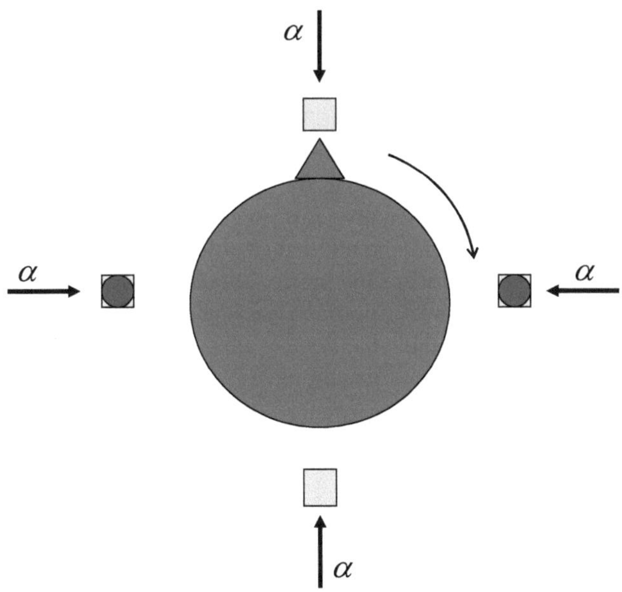

Fig. 5.1 Cyclic Polling System

duration of time spent by the UAV at an alert station and decreases monotonically with the delay in responding to alerts.

The alerts' arrival rate is a key parameter of the problem. In general, while the probability of an actual incursion can be quite small, nuisance trips are common and dominate the perimeter patrol operation. Significant resources can be squandered in determining that alerts are actually false alarms. At the same time, servicing the alerts in a timely manner is critical. A *distinguishing feature* of our work is the consideration, in the design of the optimal controller, of both the information gathered by the UAV, as well as the delay in responding to an alert, while also acknowledging up front the possibility of operator error, and, in particular, the possibility of false alarms. The performance of the operator in classifying alerts improves with an increase in the amount of available information about the alert and the longer a UAV dwells at an alert site, the more information it can gather and transmit to the operator - consequently, the probability of detecting an intrusion increases and the probability of false identification decreases.

The problem is formulated as a stochastic dynamic program for which the optimal stationary policy can be obtained via Dynamic Programming (DP). Unfortunately, for practical values of *m*, the problem essentially becomes intractable and hence we employ an approximate linear programming method to arrive at provably good sub-optimal policies. In the next section, we introduce the linear programming approach to DP before delving into the proposed methodology. In the past, the authors had considered both exact and approximate DP methods to solve the perimeter patrol problem and variants thereof [8, 21, 22, 23, 24].

5.2 Linear Programming Approach to Dynamic Programming

The Linear Programming (LP) approach to solving dynamic programs (DPs) originated from the papers: [28, 11, 10, 16]. The basic feature of an LP approach for solving DPs corresponding to maximization of a discounted payoff is that the optimal solution of the DP (also referred to as the optimal value function) is the optimal solution of the LP for *every* non-negative cost function. The constraint set describing the feasible solution of the LP and the number of independent variables are typically very large (*curse of dimensionality*) and hence, obtaining the exact solution of a DP (stochastic or otherwise) via an LP approach is not practical. Despite this limitation, an LP approach provides a tractable method for approximate dynamic programming [29, 35, 41] and the advantages of this approach may be summarized as follows:

1. One can restrict the value function to be of a certain parameterized form, thereby reducing the dimension of the LP to the size of the parameter set to make it tractable.
2. The solution to the LP provides upper bounds for the value function (lower bounds, if minimizing a discounted cost, as opposed to maximizing discounted payoff, is considered as the optimization criteria).

The main questions regarding the tractability and quality of approximate DP revolve around restricting the value function in a suitable way. The questions are: (1) How does one restrict the value function, i.e., what basis functions should one choose for parameterizing the value function? (2) Are there any (a posteriori) bounds that one can provide about the value function from the solution of a restricted LP? If the restrictions imposed on the value function are consistent with the physics/structure of the problem, one can expect reasonably tight bounds. There is another question that naturally arises: In the unrestricted case, the optimal solution of the LP is independent of the choice of the non-negative cost function. While it is unreasonable to expect that the optimal value function be a feasible solution of the restricted LP, one can ask if the optimal solution of the restricted LP is the same for *every* choice of non-negative cost function for the LP. It has been reported in the literature that this is unfortunately not the case [9].

If the LP is not properly restricted, it can lead to poor approximation and perhaps, even infeasibility [12]. A common approach is to approximate the value (cost-to-go) function by a linear functional of a priori chosen basis functions [35]. This approach is attractive in that for a certain class of basis functions, feasibility of the approximate (or restricted) LP is guaranteed [9]. A straightforward method for selecting the basis functions is through a state aggregation method. Here the state space is partitioned into disjoint sets or partitions and the approximate value function is restricted to be the same for all the states in a partition. The number of variables for the LP therefore reduces to the number of partitions. State aggregation based approximation techniques were originally proposed by [2, 3, 30]. Since then, substantial work has been reported in the literature on this topic (see [42] and the reference therein). In this work, we adopt the state aggregation method.

Although imposing restrictions on the value function reduces the size of the restricted LP, the number of constraints does not change. Since the number of constraints is at least of the same order as the number of states of the DP, one is faced with a restricted LP with a large number of constraints. An LP with a large number of constraints may be solved if there is an automatic way to separate a non-optimal solution from an optimal one [14]; otherwise, one may have to resort to heuristics or settle for an approximate solution. Separation of a non-optimal solution from an optimal one is easier if one has a compact representation of constraints [31] or if a subset of the constraints that dominate other constraints can easily be identified from the structure of the problem [23]. Popular methods to prune the constraint set include aggregation of constraints, sub-sampling of constraints [9], constraint generation methods [13, 34] and other heuristic methods [40].

If the solution of the restricted LP is the same for *every* non-negative cost function of the LP, then it suggests that the constraint set for the restricted LP embeds the constraint set for the exact LP corresponding to a reduced order Markov Decision Process (MDP). If one adopts a naive approach and "aggregates" every state into a separate partition, we obtain the original exact LP and clearly, for this LP, the solution is independent of the non-negative cost function. It would seem reasonable to expect that this would generalize to partitions of arbitrary size and in fact, we prove this to be the case in this article. One can construct a sub-optimal policy from the solution to the restricted LP by considering the policy that is greedy with respect to the approximate value function [33]. By construction, the expected discounted payoff for the sub-optimal policy will be a lower bound to the optimal value function and hence, can be used to quantify the quality of the sub-optimal policy. Also the lower bound will be closer to the optimal value function than the approximate value function by virtue of the monotonicity property of the Bellman operator. But the lower bound computation is not efficient since the procedure involved is tantamount to policy evaluation which involves the solution to a system of linear equations of the same size as the state-space. In this work, we have developed a novel disjunctive LP, whose solution can be used to construct a lower bound to the optimal value function. The contributions of our work may be summarized as follows:

- The solution to a state aggregation based restricted LP is shown to be *independent* of the underlying (positive) cost function. Moreover, the optimal solution is dominated by every feasible solution to the restricted LP [32].
- A subset of the constraints of the restricted LP can be used for constructing a lower bound for the optimal value function. However, this involves solving a disjunctive LP, which is (in general) not tractable.
- We demonstrate the use of aggregation based restricted LPs for the perimeter patrol scenario. For the application considered, we show that both the lower bounding disjunctive LP and the upper bounding restricted LP can be solved efficiently since they both reduce to exact LPs corresponding to some lower dimensional MDPs.

The rest of the chapter is organized as follows: we set up a mathematical model for the perimeter patrol problem in Sect. 5.3. The problem is then posed as a Markov decision process in Sect. 5.4. We introduce the state aggregation method and the restricted LP approach in Sect. 5.5. Some general results on the restricted upper bounding LP are shown in Sect. 5.5.1 followed by detailed enumeration of the structure specific to the patrol problem in Sect. 5.5.2. A novel disjunctive LP that provides a lower bound on the optimal value function is developed in Sect. 5.5.3, again followed by its efficient characterization for the patrol problem. Finally, we provide supporting simulation results in Sect. 5.6, followed by the summary in Sect. 5.7.

5.3 Perimeter Patrol: Problem Statement

The patrolled perimeter is a simple closed curve with $N(\geq m)$ nodes which are (spatially) uniformly separated, of which m correspond to the alert stations. Let the m distinct station locations be elements of the set $\Omega \subset \{0,\dots,N-1\}$. A typical scenario shown in Fig. 5.2 has 15 nodes, of which, nodes $\{0,3,7,11\}$ correspond to the alert stations. Here, station locations 3, 7 and 11 have no alerts, and station location 0 has an alert being serviced by the UAV. At time instant t, let $\ell(t)$ be the position of the UAV on the perimeter ($\ell \in \{0,\dots,N-1\}$), $d(t)$ be the dwell time (number of loiters completed if at an alert site) and $\tau_j(t)$ be the delay in servicing an alert at location $j \in \Omega$. Let $y_j(t)$ be a binary, but random, variable indicating the arrival of an alert at location $j \in \Omega$. We will assume that the statistics associated with the random variable $y_j(t)$ are known and that $y_j; j \in \Omega$ are independent. The alert arrival process is modeled as follows: Each station has an independent Poisson arrival stream of alerts at the rate of α (alerts per unit time). Once a station has an alert waiting, no new alerts can arrive there until the current one is serviced. Hence, there are 2^m possibilities for the vector of alerts, $y(t) = [y_1(t)\, y_2(t)\, \cdots y_m(t)]$ ranging from the binary equivalent of 0 to $2^m - 1$, indicating whether or not there is an alert, waiting to be serviced, at each of the m stations.

The control decisions are indicated by the variable u. If $u = 1$, then the UAV continues in the same direction as before; if $u = -1$, then the UAV reverses its direction of travel and if $u = 0$, the UAV dwells at the current alert station. We will assume that a UAV advances by one node in unit time, if $u \neq 0$. We also assume that the time to complete one loiter is also the unit time. We denote the UAV's direction of travel by ω, where $\omega = 1$ and $\omega = -1$ indicate the clockwise and counter-clockwise directions respectively. One may write the state update equations for the system as follows:

$$
\begin{aligned}
\ell(t+1) &= [\ell(t) + \omega(t)u(t)] \bmod N, \\
\omega(t+1) &= \omega(t)u(t) + \delta(u(t)), \\
d(t+1) &= (d(t)+1)\delta(u(t)), \qquad\qquad (5.1) \\
\tau_j(t+1) &= (\tau_j(t)+1)\{(1-\delta(\ell(t)-j)\delta(u(t))\}\max\{\sigma(\tau_j(t)), y_j(t)\}, \quad \forall j \in \Omega,
\end{aligned}
$$

where δ is the Kronecker delta function and $\sigma(\cdot) = 1 - \delta(\cdot)$. We denote the status of the alert at station location $j \in \Omega$ at time t by $\mathscr{A}_j(t)$, i.e.,

$$
\mathscr{A}_j(t) = \begin{cases} 0, & \text{if } \tau_j(t) = 0 \\ 1, & \text{otherwise} \end{cases}, \forall j \in \Omega. \qquad (5.2)
$$

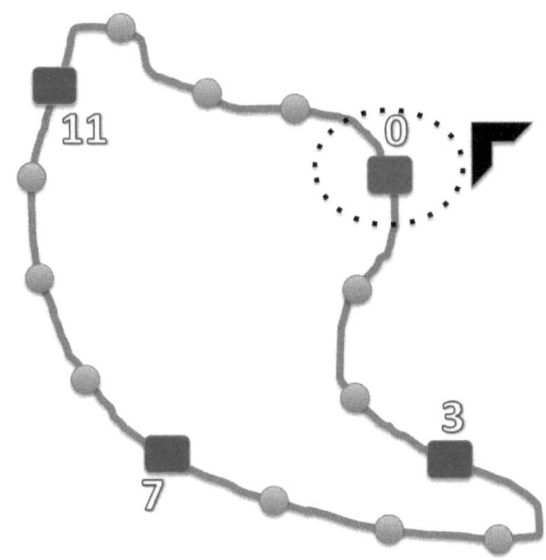

Fig. 5.2 Perimeter patrol scenario with UAV servicing alert at station 0

Also, we have the constraints: $u(t) = 0$ only if $\ell(t) \in \Omega$ and $d(t) \leq D$. If $d(t) = D$, then $u(t) \neq 0$ i.e., the UAV is forced to leave the station if it has already completed the maximum (allowed) number of dwell orbits. Combining the different components in (5.1), we express the evolution equations compactly as:

$$\mathbf{x}(t+1) = f(\mathbf{x}(t), u(t), y(t)), \tag{5.3}$$

where, $\mathbf{x}(t)$ is the system state at time t with components $\ell(t)$, $\omega(t)$,$d(t)$ and $\tau_j(t)$, $\forall j \in \Omega$. Let us denote the 2^m possible values (from the m digit binary representation of 0 to $2^m - 1$) that $y(t)$ can take by the row vector $\tilde{y}_j \in \mathscr{R}^m, j = 1, \ldots, 2^m$. Given that the alert arrival process is Poisson with parameter α, the probability that there is no alert in a unit time step, $q = e^{-\alpha}$ and hence, the probability that $y(k)$ takes any one of 2^m possible values is given by,

$$p_j := \mathscr{P}\{y(t) = \tilde{y}_j\} = q^{(m-a_j)}(1-q)^{a_j}, \quad j \in \{1, \ldots, 2^m\}, \tag{5.4}$$

where $a_j = \sum_{i=1}^{m} \tilde{y}_j(i)$ denotes the number of stations with alerts for the alert arrival configuration indicated by $\tilde{y}_j$.

5.4 Markov Decision Process

Let $\mathscr{S}$ denote the set of all possible system states and henceforth, we use x to denote a particular state $x \in \{1, \ldots, |\mathscr{S}|\}$. From the state definition, we compute the total number of states to be,

$$|\mathscr{S}| = 2 \times N \times (\Gamma + 1)^m + D \times m \times (\Gamma + 1)^{m-1}, \tag{5.5}$$

where, the factor 2 comes from the UAV being bi-directional. For the loiter states, directionality is irrelevant and hence when $d \geq 1$, we reset ω to be 1. Note that, in lieu of the reward function definition (5.6), we do not keep track of delays beyond Γ and hence the state-space $\mathscr{S}$ only includes states x with $\tau_i \leq \Gamma, \forall i \in \Omega$ and so, is finite.

Our objective is to find a suitable policy that simultaneously minimizes the service delay and maximizes the information gained upon loitering. The information gain, $\mathscr{I}$, which is based on an operator error model (see Appendix for details), is plotted as a function of dwell time in Fig. 5.3. We model the one-step payoff/ reward function as follows:

$$R_u(x) = [\mathscr{I}(d_x + 1) - \mathscr{I}(d_x)]\,\delta(u) - \rho \max\{\bar{\tau}_x, \Gamma\}, \quad x = 1, \ldots, |\mathscr{S}|, \tag{5.6}$$

where d_x is the dwell associated with state x and $\bar{\tau}_x = \max_{j \in \Omega} \tau_{j,x}$ is the worst service delay (among all stations) associated with state x. The parameter $\Gamma (>> 0)$ is a judiciously chosen maximum penalty. The positive parameter ρ is a constant weighing the incremental information gained upon loitering once more at the current location against the delay in servicing alerts at other stations.

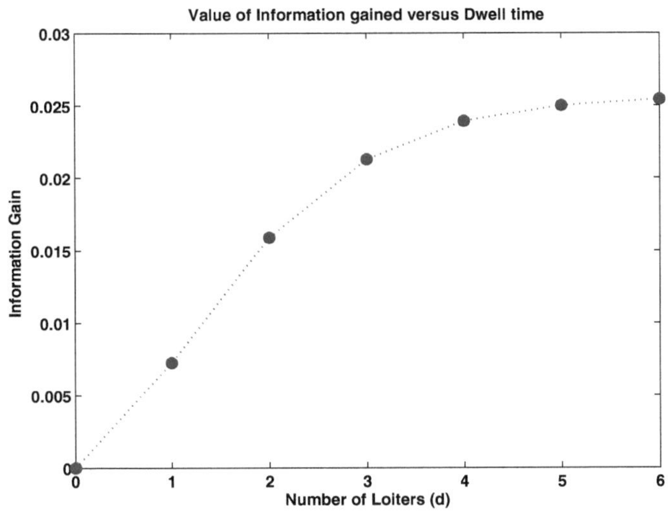

Fig. 5.3 Information gain as a function of dwell time

Any *stationary* policy, π, specifies for each state $x \in \{1,\ldots,|\mathscr{S}|\}$, a control action $u = \pi(x)$. We write the transition probability matrix for a fixed u to be P_u, where $P_u(i,j)$ indicates the probability of transitioning from state i to state j under the influence of u (the states are ordered arbitrarily). From the Poisson distribution (5.4), we have:

$$P_u(x,y) = \begin{cases} 0, \text{ if } y \neq f(x,u,\tilde{y}_l) \text{ for any } l \in \{1,\ldots,2^m\}, \\ \sum_{j\in\mathscr{C}} p_j, \text{ where } \mathscr{C} = \{l | y = f(x,u,\tilde{y}_l)\}. \end{cases} \tag{5.7}$$

We express the column vector of immediate payoffs associated with the policy π to be R_π, where $R_\pi(x) = R_{\pi(x)}(x)$. We are interested in solving a stochastic control problem, which amounts to selecting a policy that maximizes the infinite-horizon discounted reward starting from x_0,

$$V_\pi(x_0) = \mathbf{E}\left[\sum_{t=0}^{\infty} \lambda^t R_\pi(x(t)) \middle| x(0) = x_0\right], \tag{5.8}$$

where $\lambda \in [0,1)$ is a temporal discount factor. We obtain the optimal policy by solving Bellman's equation,

$$V^*(x) = \max_u \left\{ R_u(x) + \lambda \sum_{l=1}^{2^m} p_l V^*(f(x,u,\tilde{y}_l)) \right\}, \forall x \in \mathscr{S}, \tag{5.9}$$

where, $V^*(x)$ is the optimal value function (or optimal discounted payoff) starting from state x. The optimal policy then is given by,

$$\pi^*(x) = \arg\max_u \left\{ R_u(x) + \lambda \sum_{l=1}^{2^m} p_l V^*(f(x,u,\tilde{y}_l)) \right\}, \forall x \in \mathscr{S}. \tag{5.10}$$

Recall that the problem size, $|\mathscr{S}|$, is an mth order polynomial in Γ (5.5). So, solving for the optimal value function and policy using exact dynamic programming (DP) methods [4, 17] are rendered intractable for practical values of Γ and m. For this reason, one is interested in tractable approximate methods that yield suboptimal solutions with some guarantees on the deviation of the associated approximate value function from the optimal one. Before venturing into the approximation scheme, we introduce some preliminary results, that help us in establishing the aggregation based LPs.

5.4.1 Linear Programming Preliminaries

Bellman's equation suggests that the optimal value function satisfies the following set of linear inequalities, which we will refer to as the Bellman inequalities:

$$V(x) \geq R_u(x) + \lambda \sum_{l=1}^{2^m} p_l V(f(x,u,\tilde{y}_l)), \ \forall u, \ \forall x \in \mathscr{S}. \tag{5.11}$$

The Bellman inequalities may be compactly represented as:

$$V \geq R_u + \lambda P_u V, \ \forall u. \tag{5.12}$$

It is a well known result that any V that satisfies the Bellman inequalities is an upper bound to the optimal value function V^*.

Lemma 1. *For any V that satisfies (5.12), we have $V \geq V^*$.*

Proof. For every stationary policy π, we have:

$$\begin{aligned} V &\geq R_\pi + \lambda P_\pi V, \\ \Rightarrow [I - \lambda P_\pi] V &\geq R_\pi. \end{aligned} \tag{5.13}$$

Since P_π is a stochastic matrix (i.e., it is non-negative and its row sum equals 1), and $\lambda \in [0,1)$, the matrix $[I - \lambda P_\pi]^{-1}$ admits the following analytic series expansion:

$$[I - \lambda P_\pi]^{-1} = I + \lambda P_\pi + \lambda^2 P_\pi^2 + \ldots. \tag{5.14}$$

So, all the entries of $[I - \lambda P_\pi]^{-1}$ are non-negative and hence (5.13) implies the following (although the converse is not true!):

$$V \geq [I - \lambda P_\pi]^{-1} R_\pi = \sum_{i=0}^{\infty} \lambda^i P_\pi^i R_\pi, \quad \forall \pi. \tag{5.15}$$

So, V dominates the expected payoff associated with every policy π, including the optimal policy π^*. Hence $V \geq V^*$. □

The above property leads to the following well known result in approximate DP.

Corollary 1. *The optimal solution to the so-called "exact LP":*

$$\begin{aligned} ELP &:= \min c^T V, \quad \textit{subject to} \\ V &\geq R_u + \lambda P_u V, \quad \forall u, \end{aligned} \tag{5.16}$$

is the optimal value function, V^*, for every $c > 0$.

By definition (5.9), V^* is a feasible solution to *ELP*. From Lemma 1, we have that every feasible solution, $V \geq V^* \Rightarrow c^T(V - V^*) \geq 0$. Hence, V^* is the unique optimal solution to *ELP* for every $c > 0$.

5.5 State Aggregation Based Approximate DP

In the perimeter patrol problem considered herein, we see that, by definition (5.6), there is a structure to the reward function, $R_u(x)$. To explain this structure, consider a station where an alert is being serviced by the UAV. The information gained by the UAV about the alert is only a function of the service delay at the station and the amount of time the UAV dwells at the station servicing the alert. So, no matter what the delays are at the other stations, the reward is the same, as long as the maximum delay and the dwell time of the UAV at the station are the same. This leads to a natural partitioning of the state-space, in that, we aggregate all the states that have the same ℓ, ω, d, $\mathscr{A}_j$, $\forall j \in \Omega$ and $\bar{\tau} = \max_{j \in \Omega} \tau_j$, into one partition. As a result of aggregation, the number of partitions can be shown to be,

$$M = 2 \times N + 2 \times N \times (2^m - 1) \times \Gamma + m \times D + m \times D \times (2^{m-1} - 1) \times \Gamma, \tag{5.17}$$

which is linear in Γ and hence considerably smaller than the total number of states (5.5). As per the above scheme, let the state-space $\mathscr{S}$ be partitioned into M disjoint sets, $\mathscr{S}_i, i = 1, \ldots, M$. Henceforth, we will use the following notation: if $f(x,u,Y)$ represents the state the system transitions to starting from x and subject to a control input u and a stochastic disturbance Y, then $\bar{f}(x,u,Y)$ represents the partition to which the final state belongs. We now establish the approximate LP method, defined over the space of partitions, in the following section.

5.5.1 Restricted Linear Program

Let us restrict the exact LP (5.16) by requiring further that $V(x) = v(i)$ for all $x \in \mathscr{S}_i$, $i = 1, \ldots, M$. Augmenting these constraints to the exact LP, one gets the following restricted LP, for some $c > 0$.

$$RLP := \min \sum_{i=1}^{M} \underbrace{\sum_{x \in \mathscr{S}_i} c(x)}_{\bar{c}(i)} v(i) \quad \text{subject to} \tag{5.18}$$

$$v(i) \geq R_u(x) + \lambda \sum_{l=1}^{2^m} p_l v(\bar{f}(x,u,\tilde{y}_l)), \; \forall x \in \mathscr{S}_i,$$
$$i = 1, \ldots, M, \; \forall u.$$

The restricted LP can also be written in the following compact form:

$$RLP = \min c^T \Phi v \quad \text{subject to} \tag{5.19}$$
$$\Phi v \geq R_u + \lambda P_u \Phi v, \quad \forall u,$$

where the columns of Φ (commonly referred to as "basis functions", in the literature) are given by,

$$\Phi(x,i) = \begin{cases} 1, \text{ if } x \in \mathscr{S}_i, \\ 0, \text{ otherwise.} \end{cases}, \quad i = 1, \ldots, M. \tag{5.20}$$

The restricted LP deals with a smaller number of variables, $M(<< |\mathscr{S}|)$. An approximate value function can be constructed from every feasible solution to *RLP* according to $\tilde{V} = \Phi v \Leftrightarrow V(x) = v(i), \; \forall x \in \mathscr{S}_i, i = 1, \ldots, M$. Since the approximate value function satisfies, by construction, the Bellman inequalities (5.11), it is automatically an upper bound to V^* (by Lemma 1). So, if v_c^* is the optimal solution to *RLP* (5.18) for some cost vector $\bar{c} > 0$, then clearly, $v_c^*(i) \geq V^*(x), \; \forall x \in \mathscr{S}_i, i = 1, \ldots, M$. We now establish a general result on the optimal solution to the aggregation based restricted LP.

Lemma 2. *If w is a feasible solution to RLP, then $w \geq v_c^*$, where v_c^* is the optimal solution to RLP for some $\bar{c} > 0$.*

Proof. Let z to be the element-wise minimum of v_c^* and w (both are bounded from below), i.e., $z(i) = \min(v_c^*(i), w(i)), \; i = 1, \ldots, M$. Since both v_c^* and w satisfy the Bellman inequalities (constraints of *RLP*), we necessarily have,

$$z(i) \geq \left\{ R_u(x) + \lambda \sum_{l=1}^{2^m} p_l z(\bar{f}(x,u,\tilde{y}_l)) \right\}, \; \forall x \in \mathscr{S}_i,$$
$$i = 1, \ldots, M, \; \forall u. \tag{5.21}$$

So, z is also a feasible solution to *RLP*. Suppose $\exists \; j \in \{1, \ldots, M\}$ such that $w(j) < v_c^*(j)$. Then, by definition, $z \leq v_c^*$ and furthermore, $z(j) < v_c^*(j)$. This leads to $\sum_{i=1}^{M} \bar{c}(i) z(i) < \sum_{i=1}^{M} \bar{c}(i) v_c^*(i)$, which is a contradiction, since v_c^* is the optimal solution with minimum cost! So, we conclude that $w \geq v_c^*$. □

Corollary 2. *The optimal solution, v^*, to the RLP is independent of the underlying positive cost vector $\bar{c}$.*

Proof. If v_1^* is the optimal solution to *RLP* with cost vector $\bar{c}_1 > 0$ and v_2^* is the optimal solution to *RLP* with a different cost vector $\bar{c}_2 > 0$, then it immediately follows from Lemma 2 that they necessarily dominate each other; and hence, $v_1^* = v_2^*$. □

Lemma 2 implies that the upper bound for the optimal value function cannot be improved by changing the cost function from a linear to a non-linear function or by restricting the feasible set of *RLP* further since the optimal solution is dominated by every other feasible solution. Also Φv^* is the least upper bound to the optimal value function V^* since any other feasible v satisfies $v \geq v^* \Rightarrow \Phi v \geq \Phi v^*$.

Given that the choice of cost function has no bearing on the *RLP* solution, we now focus on how to solve it for the perimeter patrol application. As was mentioned earlier, although the *RLP* deals with a small number of variables, M, the number of constraints is the same as the exact LP (5.16). So, solving the *RLP* is no less difficult than solving the original MDP itself! Recall the *RLP* constraints for every partition $i = 1, \ldots, M$ and control u,

$$\begin{aligned} v(i) &\geq R_u(x) + \lambda \sum_{l=1}^{2^m} p_l v(\bar{f}(x,u,\tilde{y}_l)), \ \forall x \in \mathscr{S}_i, \\ \Rightarrow v(i) &\geq \max_{x \in \mathscr{S}_i} \left\{ R_u(x) + \lambda \sum_{l=1}^{2^m} p_l v(\bar{f}(x,u,\tilde{y}_l)) \right\}. \end{aligned} \tag{5.22}$$

Luckily, structure in the perimeter patrol problem enables one to identify, a priori, which $x \in \mathscr{S}_i$ corresponds to the dominating (or binding) constraint; thereby providing an efficient (and tractable) characterization of the *RLP*. We demonstrate this central result in the following section.

5.5.2 *Partial Ordering of States*

Let $\ell_x, d_x, \omega_x, \tau_{j,x}$ and $\mathscr{A}_{j,x}$ represent respectively, the location, dwell, direction of UAV's motion and the service delay and alert status at station location $j \in \Omega$ corresponding to some state $x \in \{1, \ldots, |\mathscr{S}|\}$. Also, we will use $\ell(i), d(i), \omega(i)$, $\bar{\tau}(i)$ and $\mathscr{A}_j(i)$ to denote the location, dwell, direction, maximum delay (among all stations), and the alert status at station location $j \in \Omega$ that correspond to partition index $i \in \{1, \ldots, M\}$. We also denote by $x(t; x_0, \mathbf{u}_t, \mathbf{y}_t)$ the state at time $t > 0$; if the initial state at $t = 0$ is x_0 and the sequence of inputs, $\mathbf{u}_t = \{u(0), u(1), \ldots, u(t-1)\}$ and disturbances, $\mathbf{y}_t = \{y(0), y(1), \ldots, y(t-1)\}$. We introduce a partial ordering of the states according to: $x \geq y$ iff $\ell_x = \ell_y$, $d_x = d_y$, $\omega_x = \omega_y$ and $\tau_{j,x} \geq \tau_{j,y}$, $\forall j \in \Omega$. By the same token, we also partially order partitions: $\mathscr{S}_i \geq \mathscr{S}_j$ iff for every $z \in \mathscr{S}_j$, there $\exists\, x \in \mathscr{S}_i$ such that $x \geq z$.

Definition 1. From a given state $x \in \mathscr{S}_i$, we define $z_x^{i,u}$ to be the tuple of partition indices that the system can transition to, under control action u i.e., $z_x^{i,u} = (\bar{f}(x,u,\tilde{y}_1), \ldots, \bar{f}(x,u,\tilde{y}_{2^m}))$. We denote by $\mathscr{T}(i,u)$ the set of all distinct $z_x^{i,u}$ for a given partition index i and control u.

Given the partial ordering of states, we have the following monotonicity result, on the state evolution.

Lemma 3. *If $x_1 \geq x_2$, then for the same sequence of inputs $\mathbf{u}_t$ and disturbances $\mathbf{y}_t$, we have $x(t;x_1,\mathbf{u}_t,\mathbf{y}_t) \geq x(t;x_2,\mathbf{u}_t,\mathbf{y}_t)$ for every $t > 0$.*

Proof. We use induction. Clearly at $t = 0$, $x_1 \geq x_2$. By the semi-group property of state transitions, it is sufficient to show that the result holds for $t = 1$. We define the state, x, of the patrol system to be of two types.

Type 1 If the UAV is at a station with an alert, the dwell time is zero and there is an alert at some other station, i.e.,

$$\ell_x \in \Omega,\ d_x = 0,\ \mathscr{A}_{\ell_x,x} = 1,\ \text{and}\ \mathscr{A}_{j,x} = 1,\ \text{for some}\ j \in \Omega, j \neq \ell_x. \tag{5.23}$$

Else, it is of Type 2.

Note that if $x_1 \geq x_2$, then the states x_1 and x_2 are necessarily of the same type. The key property we will be using in proving this result, is the following: service delay at a station either remains at zero (if no new alert has occurred there) or it goes up by 1 (if there is an unserviced alert there) or it is reset to zero (if a UAV decides to loiter there). If x_1 and x_2 are of Type 1 and the UAV chooses to loiter, i.e., $u(0) = 0$, we clearly see that neither the location nor the dwell will differ at $t = 1$. Furthermore, the delays at $t = 1$ associated with the stations corresponding to initial state x_1 will be no less than the delays associated with stations corresponding to initial state x_2 since $x_1 \geq x_2$. If $z_1 = x(1;x_1,0,y(0))$ and $z_2 = x(1;x_2,0,y(0))$, we see that $\ell_{z_1} = \ell_{z_2}$, $d_{z_1} = d_{z_2}$, $\omega_{z_1} = \omega_{z_2}$, and $\tau_{j,z_1} \geq \tau_{j,z_2}$, $\forall j \in \Omega$ for every disturbance $y(0)$ and so $z_1 \geq z_2$. The same relationship holds for other possible control choices, $u(0) \neq 0$, as well. By a similar argument, one can show that $x(1;x_1,u(0),y(0)) \geq x(1,x_2,u(0),y(0))$ holds, regardless of the control choice, even if the states x_1, x_2 are of Type 2. □

The partial ordering and the monotonicity result bring about a useful property in the optimal value function, that is central to the *RLP* simplification; which we state here.

Lemma 4. *If $x_1 \geq x_2$, then $V^*(x_1) \leq V^*(x_2)$. Furthermore, if $\mathscr{S}_i \geq \mathscr{S}_j$, then $\min_{x\in\mathscr{S}_i} V^*(x) \leq \min_{z\in\mathscr{S}_j} V^*(z)$.*

Proof. Let π^* be the optimal policy; accordingly $\pi^*(x)$ is fixed for every $x \in \mathscr{S}$. Then, for every $t > 0$, we can determine $x(t;x_1,\mathbf{u}_t^*,\mathbf{y}_t)$ for some sequence of disturbances $\mathbf{y}_t$, where the optimal input sequence $\mathbf{u}_t^* = \{u^*(0),\ldots,u^*(t-1)\}$ (starting with x_1) can be recursively obtained as follows:

$$u^*(t) = \pi^*(x(t-1;x_1,\mathbf{u}_{t-1}^*,\mathbf{y}_{t-1})), \tag{5.24}$$

with the initialization $u^*(0) = \pi^*(x_1)$. For the above $\mathbf{u}^*$ and $\mathbf{y}$, we can then determine the evolution of the states corresponding to initial state x_2. Since $x(t;x_1,\mathbf{u}_t^*,\mathbf{y}_t) \geq x(t;x_2,\mathbf{u}_t^*,\mathbf{y}_t)$ by Lemma 3, we notice readily that the reward $R_{u^*}(x(t;x_1,\mathbf{u}_t^*,\mathbf{y}_t)) \leq$

$R_{u^*}(x(t;x_2,\mathbf{u}_t^*,\mathbf{y}_t))$ for every $t \geq 0$ (since the one-step reward is based only on the maximum delay, dwell time and control input, the inequality follows). Since the above holds for any given disturbance sequence, the expected discounted payoff associated with the state starting from x_1 i.e., $V^*(x_1)$, is no more than the expected discounted payoff associated with the state starting from x_2, which we will denote by $V_{\mathbf{u}^*}(x_2)$. As a result, $V^*(x_1) \leq V_{\mathbf{u}^*}(x_2) \leq V^*(x_2)$. The second part of the inequality holds since $\mathbf{u}_t^*$ as defined in (5.24) is a sub-optimal control policy for the state evolution starting from x_2 and hence the expected discounted payoff associated with that policy is necessarily dominated by the optimal discounted payoff starting from x_2.

For the second part of the result, consider two different partitions $\mathscr{S}_i$ and $\mathscr{S}_j$ such that $\mathscr{S}_i \geq \mathscr{S}_j$. Let $\bar{z} = \arg\min_{z\in\mathscr{S}_j} V^*(z)$. Since $\mathscr{S}_i \geq \mathscr{S}_j$, $\exists\, \bar{x} \in \mathscr{S}_i$ such that $\bar{x} \geq \bar{z}$. We have shown that for this case, $V^*(\bar{x}) \leq V^*(\bar{z}) = \min_{z\in\mathscr{S}_j} V^*(z) \Rightarrow \min_{x\in\mathscr{S}_i} V^*(x) \leq \min_{z\in\mathscr{S}_j} V^*(z)$ □

Before embarking on the *RLP* simplification, we establish an intermediate result, that we need, on the size of the transition map $\mathscr{T}(i,u)$. For the sake of notational simplicity, we denote the lth component of any tuple $k \in \mathscr{T}(i,u)$ by k_l and the cardinality of the set $\mathscr{T}(i,u)$ by $|\mathscr{T}(i,u)|$. Also analogous to the states, we define the partitions to be of two types:

Type 1 If the UAV is at a station with an alert, the dwell time is zero and there is an alert at some other station, i.e.,

$$\ell(i) \in \Omega,\ d(i) = 0,\ \mathscr{A}_{\ell(i)}(i) = 1, \text{ and } \mathscr{A}_j(i) = 1, \text{ for some } j \in \Omega, j \neq \ell(i), \tag{5.25}$$

Else, it is of Type 2. If a partition index i is of Type 1, we say $i \in \mathscr{P}_1$. Given this definition, we have the following result on the cardinality of $\mathscr{T}$:

Lemma 5

$$|\mathscr{T}(i,u)| = \begin{cases} \bar{\tau}(i),\ i \in \mathscr{P}_1 \textit{ and } u = 0, \\ 1,\ \textit{otherwise.} \end{cases}$$

Proof. First we consider partition index $i \in \mathscr{P}_1$ and control input $u = 0$. Since the UAV has decided to loiter at the current station, $\ell(i) \in \Omega$, the service delay at that station, $\tau_{\ell(i)}$ will be reset to zero in the next time step. Hence the future state (and partition) maximum delay will be determined by the highest of the service delays, say $\tilde{\tau}$, among the other stations with alerts (at least one such station exists since partition $i \in \mathscr{P}_1$). So $\forall j \in \{1,\ldots,\bar{\tau}(i)\}$, $\exists\, x_j \in \mathscr{S}_i$ such that $\tilde{\tau}_{x_j} = j$. The corresponding tuple of future partition indices, $z_{x_j}^{i,0} = (\bar{f}(x_j,u,\tilde{y}_1),\ldots,\bar{f}(x_j,u,\tilde{y}_{2^m}))$ will have maximum delay $j+1$ and so, $\mathscr{T}(i,0) = \cup_{j=1}^{\bar{\tau}(i)}\{z_{x_j}^{i,0}\} \Rightarrow |\mathscr{T}(i,0)| = \bar{\tau}(i)$. For all other control choices, $u \neq 0$, all the states $x \in \mathscr{S}_i$ will transition to future states with the same maximum delay $\bar{\tau}(i)+1$. So, for $u \neq 0$, $\mathscr{T}(i,u)$ is a singleton set and hence $|\mathscr{T}(i,u)| = 1$. For partition indices j of Type 2 with $\bar{\tau}(j) > 0$, all the states $x \in \mathscr{S}_j$ will transition to future states with the same maximum delay $\bar{\tau}(j)+1$ and so, $|\mathscr{T}(j,u)| = 1$, $\forall u$. If $\bar{\tau}(j) = 0$, then the partition

$\mathscr{S}_j$ is a singleton set as per the aggregation scheme (see Sec 5.5.2) and hence $|\mathscr{T}(j,u)| = 1, \forall u$. □

We now have all the tools necessary to show that the upper bound formulation, *RLP* (5.18), collapses to an exact LP (5.16) corresponding to some lower order MDP and we do so via the following result.

Theorem 1. *For the perimeter patrol problem, the RLP (5.18) reduces to the following LP.*

$$UBLP := \min \bar{c}^T w, \quad \textit{subject to} \tag{5.26}$$
$$w(i) \geq r_u(i) + \lambda \sum_{l=1}^{2^m} p_l w(k_l), \; \forall u, \; i = 1, \ldots, M,$$

where the tuple k is the unique element in $\mathscr{T}(i,u)$ *if* $|\mathscr{T}(i,u)| = 1$. *Else,* $k = k^*$, *where* $k^* \in \mathscr{T}(i,u)$ *is the tuple of partition indices such that* $\bar{\tau}(k_l^*) = 2, \; l = 1, \ldots, 2^m$.

Proof. Given the definition of $\mathscr{T}(i,u)$ (see Def. 1), the constraints in *RLP* (5.18) can be rewritten as follows:

$$v(i) \geq \left\{ r_u(i) + \lambda \sum_{l=1}^{2^m} p_l v(k_l) \right\}, \; \forall k \in \mathscr{T}(i,u), \forall u, \tag{5.27}$$

where $r_u(i) = R_u(x), \; \forall x \in \mathscr{S}_i$ is the reward associated with partition index i. A sufficient condition for $v(i) > V^*(x), \; \forall x \in \mathscr{S}_i$ i.e., for v to be an upper bound to the optimal value function (of all states) in partition i, is the following:

$$v(i) \geq r_u(i) + \lambda \sum_{l=1}^{2^m} p_l V^*(f(x,u,\tilde{y}_l)), \; \forall u, \; \forall x \in \mathscr{S}_i. \tag{5.28}$$

For partition index $i \in \mathscr{P}_1$ and $u = 0$, this collapses to,

$$v(i) \geq r_0(i) + \lambda \sum_{l=1}^{2^m} p_l V^*(f(\bar{x},0,\tilde{y}_l)), \tag{5.29}$$

where $\bar{x} \in \mathscr{S}_i$ is the state that transitions to future states with the least maximum delay, 2. This is because, $f(\bar{x},0,\tilde{y}_l) \leq f(x,0,\tilde{y}_l), \; l = 1, \ldots, 2^m, \; \forall x \in \mathscr{S}_i$ and so we have (from Lemma 4), $V^*(f(\bar{x},0,\tilde{y}_l)) \geq V^*(f(x,0,\tilde{y}_l)), \; l = 1, \ldots, 2^m, \; \forall x \in \mathscr{S}_i$. The inequality above (5.29) suggests that the $\bar{\tau}(i)$ constraints (we know the number of constraints from Lemma 5) in *RLP* can be replaced by the single constraint,

$$v(i) \geq \left\{ r_0(i) + \lambda \sum_{l=1}^{2^m} p_l v(k_l^*) \right\}, \tag{5.30}$$

where $k^* = (\bar{f}(\bar{x},0,\tilde{y}_1), \ldots, \bar{f}(\bar{x},0,\tilde{y}_{2^m}))$ is the tuple of future partition indices (corresponding to $\bar{x}$) with the least possible maximum delay, i.e., $\bar{\tau}(k_l^*) = 2, \; l = 1, \ldots, 2^m$.

For the other control choices, $u \neq 0$, there exists only one tuple $\bar{k}$ in $\mathscr{T}(i,u)$ (since $|\mathscr{T}(i,u)| = 1$ from Lemma 5) and hence the corresponding constraints (5.27) collapse to the single constraint,

$$v(i) \geq r_u(i) + \lambda \sum_{l=1}^{2^m} p_l v(\bar{k}_l),\ u \neq 0. \tag{5.31}$$

Similarly, for partitions $\mathscr{S}_j$ of Type 2, $|\mathscr{T}(j,u)| = 1$, $\forall u$ from Lemma 5, and so the corresponding constraints (5.27) collapse to the single constraint:

$$v(j) \geq r_u(j) + \lambda \sum_{l=1}^{2^m} p_l v(\bar{k}_l),\ \forall u. \tag{5.32}$$

In summary, we have the following: regardless of which partition index $i \in \{1,\ldots,M\}$ and control action u are considered, the corresponding $|\mathscr{T}(i,u)|$ linear constraints in *RLP* collapse to a single constraint and hence, *RLP* for the perimeter patrol problem reduces to the following LP:

$$\begin{aligned} UBLP := \min \bar{c}^T w, &\quad \text{subject to} \\ w(i) \ \geq \ r_u(i) + \lambda \sum_{l=1}^{2^m} p_l w(k_l),&\ \forall u,\ i = 1,\ldots,M, \end{aligned} \tag{5.33}$$

where the tuple k is the unique element in $\mathscr{T}(i,u)$ if $|\mathscr{T}(i,u)| = 1$. Else, $k = k^*$, where $k^* \in \mathscr{T}(i,u)$ is the tuple of partition indices such that $\bar{\tau}(k^*_l) = 2,\ l = 1,\ldots,2^m$. □

Corollary 3. *UBLP is the exact LP (5.16) corresponding to a reduced order MDP defined over the space of partitions with reward vector r_u and transition probability between partitions i and j given by,*

$$\tilde{P}_u(i,j) = \begin{cases} 0, \ \textit{if}\ j \neq k_l,\ \textit{for any}\ k \in \mathscr{T}(i,u),\ l = 1,\ldots,2^m, \\ \sum_{q \in \mathscr{C}_1} p_q,\ \textit{where}\ \mathscr{C}_1 = \{l | j = k_l, k \in \mathscr{T}(i,u)\}, \\ \qquad \textit{if}\ |\mathscr{T}(i,u)| = 1, \\ \sum_{q \in \mathscr{C}_2} p_q,\ \textit{where}\ \mathscr{C}_2 = \{l | j = k^*_l, k^* \in \mathscr{T}(i,u), \\ \qquad \bar{\tau}(k^*_l) = 2,\ l = 1,\ldots,2^m\},\ \textit{otherwise}. \end{cases} \tag{5.34}$$

So, solving *UBLP* is equivalent to computing the optimal value function for the above reduced order MDP and is computationally attractive, compared to solving the original problem, since $M << |\mathscr{S}|$. Upon solving *UBLP* and obtaining the approximate value function (and upper bound) $\tilde{V} = \Phi v^*$, one can then construct the corresponding sub-optimal "greedy" policy according to:

$$\pi(x) = \arg\max_u \left\{ R_u(x) + \lambda \sum_y P_u(x,y) \tilde{V}(y) \right\}, \quad \forall x \in \mathscr{S}.$$

If we define the corresponding improvement in value function, $\alpha(x) := R_\pi(x) + \lambda \sum_y P_\pi(x,y)\tilde{V}(y) - \tilde{V}(x)$, then the following bounds hold [33, 27]:

$$\tilde{V}(x) + \frac{\min_y \alpha(y)}{1-\lambda} \leq V_\pi(x) \leq V^*(x) \leq \tilde{V}(x),\ \forall x \in \mathscr{S}, \tag{5.35}$$

where, $V_\pi = (I - \lambda P_\pi)^{-1} R_\pi$ the expected discounted payoff, corresponding to the suboptimal policy π. The above equation can be used to quantify the approximation error. In the literature, typically V_π is used as a candidate for a lower bound on the optimal value function. The reason for obtaining a lower bound is straightforward: if one obtains tight lower and upper bound approximations, then the "distance" between the two would clearly indicate the quality of the approximations. In this context, the weaker lower bound provided by (5.35) is typically very conservative. On the other hand, computation of V_π involves solving a linear system of equations of size $|\mathscr{S}|$, which is as bad as solving the original MDP. So, one is interested in an alternate efficient method to compute a lower bound. In the next section, we establish such a method and warrant its application to the perimeter patrol problem.

5.5.3 Lower Bound for the Optimal Value Function

Recall that for each $x \in \mathscr{S}_i$, $V^*(x)$ satisfies the Bellman inequality (5.11):

$$\begin{aligned} V^*(x) &\geq R_u(x) + \lambda \sum_{l=1}^{2^m} p_l V^*(f(x,u,\tilde{y}_l)), \quad \forall u, \\ &\geq R_u(x) + \lambda \sum_{l=1}^{2^m} p_l \min_{y \in \bar{f}(x,u,\tilde{y}_l)} V^*(y), \quad \forall u. \end{aligned} \tag{5.36}$$

Definition 2. Let $\bar{w}(i) := \min_{x \in \mathscr{S}_i} V^*(x),\ i = 1,\ldots,M.$

Then, it follows that,

$$\bar{w}(i) \geq \min_{x \in \mathscr{S}_i} \left\{ R_u(x) + \lambda \sum_{l=1}^{2^m} p_l \bar{w}(\bar{f}(x,u,\tilde{y}_l)) \right\}, \quad \forall u,\ i = 1,\ldots,M. \tag{5.37}$$

The above set of inequalites motivates the following non-linear program:

$$\begin{aligned} NLP &:= \min \bar{c}^T w, \quad \text{subject to} \\ w(i) &\geq \min_{x \in \mathscr{S}_i} \left\{ R_u(x) + \lambda \sum_{l=1}^{2^m} p_l w(\bar{f}(x,u,\tilde{y}_l)) \right\}, \quad \forall u,\ i = 1,\ldots,M. \end{aligned} \tag{5.38}$$

Let $w_{\bar{c}}^*$ be the optimal solution to the *NLP* for some $\bar{c} \geq 0$. By construction, we see that $\bar{w}$ (see Def. 2) is a feasible solution to the *NLP* and hence,

$$\bar{c}^T w_c^* \leq \bar{c}^T \bar{w} = \sum_{i=1}^{M} \bar{c}(i) \min_{x\in\mathscr{S}_i} V^*(x).$$

So, by choosing $\bar{c}(i) = 1$ and $\bar{c}(j) = 0$ for all $j \neq i$, one can obtain a lower bound to the optimal value function for all the states in the ith partition. Unfortunately, the *NLP* is combinatorial in nature and hence intractable, for a general MDP. However, for the perimeter patrol problem, the partial ordering property (see Sect. 5.5.2) enables one to identify, a priori, which $x \in \mathscr{S}_i$ satisfies the *min* condition in (5.38). This results in the *NLP* collapsing to an LP that can be readily solved; we demonstrate this via the following result.

Theorem 2. *For the perimeter patrol problem, the NLP (5.38) reduces to the following LP.*

$$LBLP := \min \bar{c}^T w, \quad \textit{subject to}$$

$$w(i) \geq r_u(i) + \lambda \sum_{l=1}^{2^m} p_l w(k_l), \; \forall u, \; i = 1,\ldots,M, \tag{5.39}$$

where the tuple $k \in \mathscr{T}(i,u)$, *if* $|\mathscr{T}(i,u)| = 1$, *else* $k = k^*$, *where* $k^* \in \mathscr{T}(i,u)$ *is the tuple of partition indices such that* $\bar{\tau}(k_l^*) = \bar{\tau}(i) + 1$, $l = 1,\ldots,2^m$.

Proof. Recall the non-linear constraints (5.37) satisfied by $\bar{w}(i)$ (see Def. 2) that motivated the *NLP* formulation:

$$\bar{w}(i) \geq \min_{x\in\mathscr{S}_i} \left\{ R_u(x) + \lambda \sum_{l=1}^{2^m} p_l \bar{w}(\bar{f}(x,u,\tilde{y}_l)) \right\}, \quad \forall u, \quad i = 1,\ldots,M, \tag{5.40}$$

which, given the definition of $\mathscr{T}(i,u)$ (see Def. 1), can be written in the following equivalent form:

$$\bar{w}(i) \geq r_u(i) + \lambda \min_{k\in\mathscr{T}(i,u)} \sum_{l=1}^{2^m} p_l \bar{w}(k_l), \quad \forall u, \quad i = 1,\ldots,M, \tag{5.41}$$

where $r_u(i)$ is the reward associated with partition index i, and given the partitioning scheme, satisfies $R_u(x) = r_u(i), \forall x \in \mathscr{S}_i$. Given the structure in the perimeter patrol problem, we will show that the above (5.41) will collapse to a single linear inequality constraint for every partition index i and control u. Let us focus our attention on partition index $i \in \mathscr{P}_1$ and control action $u = 0$. For this choice, the cardinality of $\mathscr{T}(i,0)$ is $\bar{\tau}(i)$ (from Lemma 5). Indeed $\exists\, \bar{x} \in \mathscr{S}_i$ such that the corresponding tuple of future partition indices $k^* = (\bar{f}(\bar{x},0,\tilde{y}_1),\ldots,\bar{f}(\bar{x},0,\tilde{y}_{2^m}))$ has the highest possible maximum delay, i.e., $\bar{\tau}(k_l^*) = \bar{\tau}(i) + 1, l = 1,\ldots,2^m$. Since $k_l^* \geq k_l$, $l = 1,\ldots,2^m$, $\forall k \in \mathscr{T}(i,u)$, we have from Lemma 4,

$$\bar{w}(k_l^*) \leq \bar{w}(k_l), \; l = 1,\ldots,2^m, \; \forall k \in \mathscr{T}(i,u).$$

So, the non-linear inequality corresponding to partition index $i \in \mathscr{P}_1$ and control $u = 0$ becomes:

$$\bar{w}(i) \geq r_0(i) + \lambda \sum_{l=1}^{2^m} p_l \bar{w}(k_l^*). \tag{5.42}$$

If $u \neq 0$, then $|\mathscr{T}(i,u)| = 1$ as per Lemma 5. So there exists exactly one tuple $\underline{\mathrm{k}}$ in $\mathscr{T}(i,u)$ and hence, the non-linear constraint (5.41) reduces to the linear inequality:

$$\bar{w}(i) \geq r_u(i) + \lambda \sum_{l=1}^{2^m} p_l \bar{w}(\underline{\mathrm{k}}_l). \tag{5.43}$$

Again, for partition indices j of type 2, $|\mathscr{T}(j,u)| = 1$, $\forall u$ as per Lemma 5. So, as before, there exists exactly one tuple $\bar{k}$ in $\mathscr{T}(j,u)$ and hence, the non-linear constraint (5.41) reduces to the linear inequality:

$$\bar{w}(j) \geq r_u(j) + \lambda \sum_{l=1}^{2^m} p_l \bar{w}(\bar{k}_l). \tag{5.44}$$

In summary, we have the following: regardless of which partition one considers, the corresponding non-linear constraint in *NLP* collapses to a linear constraint and hence, *NLP* for the perimeter patrol problem collapses to the following LP:

$$\begin{aligned} LBLP &:= \min \bar{c}^T w, \quad \text{subject to} \\ w(i) &\geq r_u(i) + \lambda \sum_{l=1}^{2^m} p_l w(k_l), \quad \forall u,\ i = 1,\dots,M, \end{aligned} \tag{5.45}$$

where the tuple $k \in \mathscr{T}(i,u)$, if $|\mathscr{T}(i,u)| = 1$, else $k = k^*$, where $k^* \in \mathscr{T}(i,u)$ is the (unique) tuple of partition indices such that $\bar{\tau}(k_l^*) = \bar{\tau}(i) + 1$, $l = 1,\dots,2^m$. □

Corollary 4. *The optimal solution, w^* to LBLP is independent of the cost $\bar{c}$ and is a lower bound to the optimal value function, V^* i.e., $\forall i \in \{1,\dots,M\}$, $w^*(i) \leq \min_{x \in \mathscr{S}_i} V^*(x)$.*

Proof. To arrive at the corollary, we observe that *LBLP* is the exact LP (5.16) corresponding to a reduced order MDP defined over the space of partitions with reward r_u and transition probability between partitions i and j given by,

$$\bar{P}_u(i,j) = \begin{cases} 0, \text{ if } j \neq k_l, \text{ for any } k \in \mathscr{T}(i,u),\ l = 1,\dots,2^m, \\ \sum_{q \in \mathscr{C}_1} p_q, \text{ where } \mathscr{C}_1 = \{l | j = k_l, k \in \mathscr{T}(i,u)\}, \\ \qquad \text{if } |\mathscr{T}(i,u)| = 1, \\ \sum_{q \in \mathscr{C}_2} p_q, \text{ where } \mathscr{C}_2 = \{l | j = k_l^*, k^* \in \mathscr{T}(i,u), \\ \qquad \bar{\tau}(k_l^*) = \bar{\tau}(i) + 1,\ l = 1,\dots,2^m\}, \text{ otherwise.} \end{cases} \tag{5.46}$$

Hence, we readily have from Lemma 1, that the optimal solution w^* is independent of $\bar{c} > 0$ and also is dominated by every feasible solution, including $\bar{w}$. Hence, $w^*(i) \leq \bar{w}(i) = \min_{x \in \mathscr{S}_i} V^*(x) \leq V^*(y)$, $\forall y \in \mathscr{S}_i$, $i = 1,\dots,M$. □

So, for the perimeter patrol problem, we have shown that both the upper and lower bound approximations to the optimal value function can be computed cheaply via the *UBLP* and *LBLP* formulations respectively. In the next section, we demonstrate the efficacy of the proposed method via simulation results.

5.6 Numerical Results

We consider a perimeter with $N = 15$ nodes of which node numbers $\{0,3,7,11\}$ are alert stations (see Fig. 5.2) and a maximum allowed dwell of $D = 5$ orbits. The other parameters were chosen to be: weighing factor, $\rho = .005$ and temporal discount factor, $\lambda = 0.9$. Based on experience, we chose the alert arrival rate $\alpha = \frac{1}{30}$. This reflects a rather low false alarm rate where we expect 1 alert to occur on average in the time taken by the UAV to complete two uninterrupted patrols around the perimeter. We set the maximum delay time, that we keep track of, to be $\Gamma = 15$; for which the total number of states comes out to be $|\mathscr{S}| = 2,048,000$. To show that the proposed approximate methodology is effective, we compute the approximate value functions via the restricted LP formulation and compare them with the optimal value function. In addition, we also compute the greedy sub-optimal policy corresponding to the approximate value function and compare it with the optimal policy in terms of the two performance metrics: alert service delay and information gained upon loitering.

We aggregate the states in the example problem based on the reward function (see section 5.5.2 for details). This results in $M = 8900$ partitions, which is considerably smaller than the original number of states, $|\mathscr{S}|$. We solve both the *UBLP* and *LBLP* formulations which give us the upper and lower bounds, $V_{up} = \Phi v^*$ and $V_{low} = \Phi w^*$ respectively, to the optimal value function V^*. Since we have the optimal value function for the example problem, we use it for comparison with the approximations. Note that for higher values of m and Γ, the problem essentially becomes intractable and one would not have access to the optimal value function. Nevertheless, one can compute v^* and w^* and the difference between the two would give an estimate of the quality of the approximation. We give a representative sample of the approximation results by choosing all the states in partitions corresponding to alert status $\mathscr{A}_j = 1, \forall j \in \Omega$ (all stations have alerts) and maximum delay $\bar{\tau} = 2$. Figure 5.4 compares the optimal value function V^* with the upper and lower bound approximate value functions, V_{up} and V_{low} for this subset of the state-space. The first 15 partitions shown in the X-axis of Fig. 5.4 i.e., partition numbers, $i = 1,\dots,15$, correspond to the clockwise states:

$$\ell = i-1, \quad d=0, \quad \omega=1, \quad \bar{\tau} = \max_{j\in\Omega} \tau_j = 2, \quad \mathscr{A}_j = 1, \forall j \in \Omega, \tag{5.47}$$

and the last 15 partitions shown in the X-axis i.e., partition numbers, $i = 16,\dots,30$, correspond to the counter-clockwise states:

$$\ell = i-N-1, \quad d=0, \quad \omega=-1, \quad \bar{\tau} = \max_{j\in\Omega} \tau_j = 2, \quad \mathscr{A}_j = 1, \forall j \in \Omega. \tag{5.48}$$

Interestingly, we notice immediately that the lower bound appears to be tighter than the upper bound. Recall that our objective is to obtain a good sub-optimal policy and so, we consider the policy that is *greedy* with respect to V_{low}:

$$\pi_s(x) = \arg\max_u \left\{ R_u(x) + \lambda \sum_{l=1}^{2^m} p_l V_{low}(f(x,u,\tilde{y}_l)) \right\}, \quad \forall x \in \{1,\ldots,|\mathscr{S}|\}. \quad (5.49)$$

To assess the quality of the sub-optimal policy, we also compute the expected discounted payoff, V_{sub} that corresponds to the sub-optimal policy π_s, by solving the system of equations:

$$(I - \lambda P_{\pi_s})V_{sub} = R_{\pi_s}. \quad (5.50)$$

Since V_{sub} corresponds to a sub-optimal policy and in lieu of the monotonicity property of the Bellman operator, the following inequalities hold:

$$V_{low} \le V_{sub} \le V^* \le V_{up}.$$

In Fig. 5.5, we compare V_{sub} with the optimal value function V^* for the clockwise states defined in (5.47) and note that the approximation is quite good. Finally, we compare the performance of the sub-optimal policy π_s with that of the optimal strategy π^* in terms of the two important metrics: service delay and information gain (measured via the dwell time). To collect the performance statistics, we ran Monte-Carlo simulations with alerts at each station generated from independent Poisson arrival streams with rate $\alpha = \frac{1}{30}$. All simulations had a run time of 30000 time units.

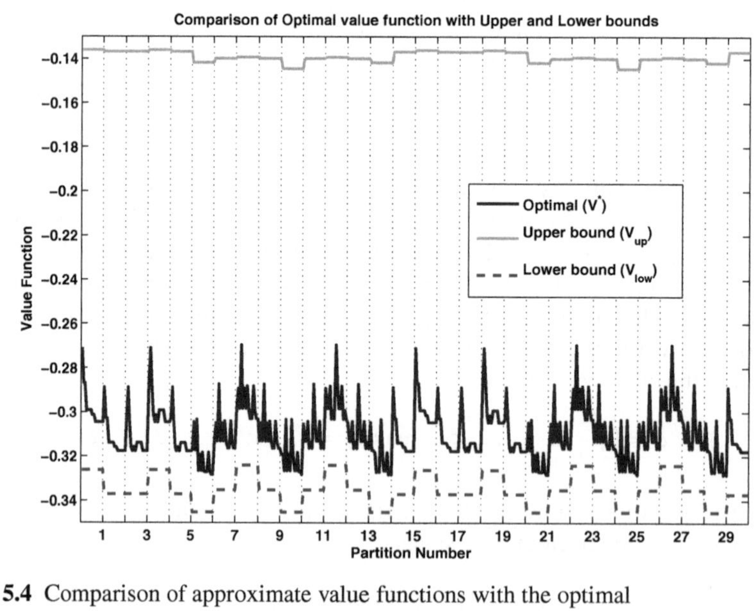

Fig. 5.4 Comparison of approximate value functions with the optimal

Both the optimal and sub-optimal policies were tested against the same alert sequence. Fig. 5.6 shows histogram plots for the service delay (top plot) and the dwell time (bottom plot) for all serviced alerts in the simulation run. The corresponding mean and worst case service delays and the mean dwell time are also shown in Table 5.1. We see that there is hardly any difference in terms of either metric between the optimal and the sub-optimal policies. This substantiates the claim that the aggregation approach gives us a sub-optimal policy that performs almost as well as the optimal policy itself. This is to be expected, given that the value functions corresponding to the optimal and sub-optimal policies are close to each other (see Fig. 5.5).

Table 5.1 Comparison of alert servicing performance between optimal and sub-optimal policies

Policy	Mean number of loiters	Mean service delay	Worst service delay
π^*	2.8	6.45	21
π_s	2.8	6.47	29

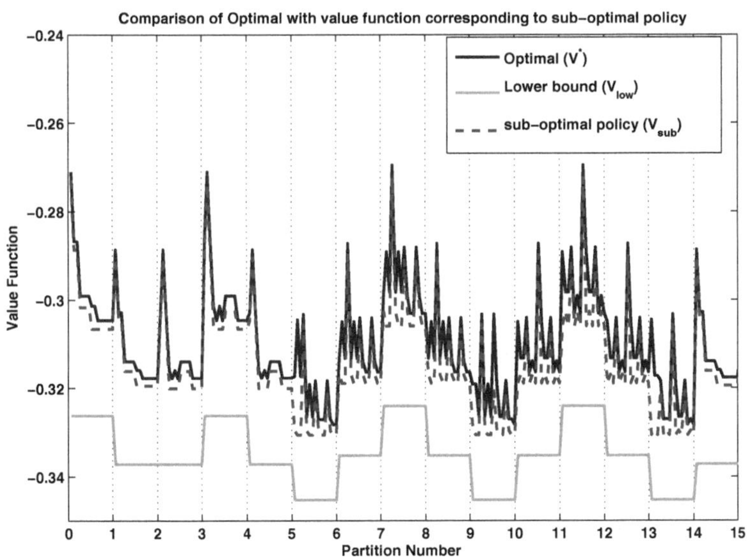

Fig. 5.5 Comparison of value function corresponding to suboptimal policy π_s with the optimal

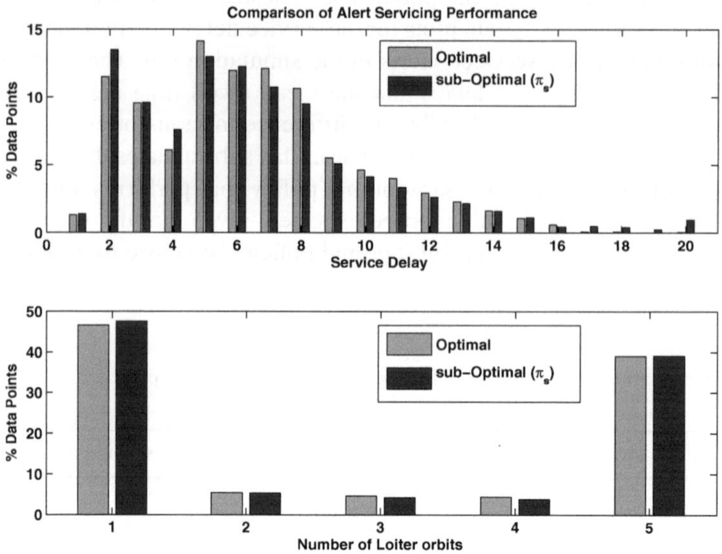

Fig. 5.6 Comparison of service delay and number of loiters between optimal and sub-optimal policies

5.7 Summary

We have provided a state aggregation based restricted LP method to construct sub-optimal policies for the perimeter patrol stochastic optimal control problem. As a general result, we have shown that the solution to an aggregation based LP is independent of the underlying positive cost function. We also provide a novel non-linear program that can be used to compute a non-trivial lower bound to the optimal value function. In particular, for the perimeter patrol problem, we have shown that both the upper and lower bound formulations simplify to exact LPs corresponding to lower dimensional MDPs defined over the space of partitions. To do so, we have exploited the partial ordering of the states that comes about because of the structure inherent in the reward function. Simulation results show that the performance of the sub-optimal policy, obtained via the lower bound approximate value function, is comparable to that of the optimal policy.

Acknowledgements. A portion of this work was completed when K. Krishnamoorthy was with the National Research Council Research Associateship program. S. Darbha was partly supported by the AFRL Summer Faculty program and AFOSR award no. FA9550-10-1-0392.

Appendix

The operator is treated as a sensor-in-the-loop automaton. To quantify the operator's performance, two random variables are considered: the variable X that specifies whether the alert is a real threat (target T) or a nuisance (false target FT) and the operator decision Z which specifies whether he determines the alert to be a real threat Z_1 or a nuisance Z_2. Let the a priori probability that an alert is a real target,

$$\mathscr{P}\{X = T\} = p << 1. \tag{5.51}$$

It is assumed, based on experience, that $p = 0.01$ in this work. The conditional probabilities which specify whether the operator correctly reported a threat and a nuisance are assumed to be functions of the discretized dwell time, $d \in \{0, \ldots, D\}$:

$$\begin{aligned} P_{TR}(d) &:= \mathscr{P}\{Z = Z_1 | X = T\} = a + b(1 - e^{-\mu_1 d}), \\ P_{FTR}(d) &:= \mathscr{P}\{Z = Z_2 | X = FT\} = c + g(1 - e^{-\mu_2 d}). \end{aligned} \tag{5.52}$$

where the acronyms TR and FTR stand for *Target Report* and *False Target Report* respectively. Together, P_{TR} and P_{FTR} determine the entries of the binary "confusion matrix" of the operator shown in Table 5.2. The parameters a, b, μ_1, c, g, μ_2 characterize the confusion matrix and the performance of the operator as a sensor (for details on sensor performance modeling, see Sec 7.2 in [18]). They satisfy the constraints:

$$0 < a + b \leq 1, \quad 0 < c + g \leq 1, \quad \mu_1 \geq 0 \quad \text{and} \quad \mu_2 \geq 0.$$

In this work, the choices $a = c = 0.5$, $b = g = 0.45$ and $\mu_1 = \mu_2 = 1$ are made. The choice $a = c = 0.5$ correspond to an uninformed or unbiased operator, i.e., the operator cannot tell if the alert is a threat or a nuisance without having seen any video footage of the alert site. Different values of a, c correspond to different types of operator bias.

Table 5.2 Operator Confusion Matrix

	Encountered Object	
Operator Decision	Target	False Target
Target	P_{TR}	$1 - P_{FTR}$
False Target	$1 - P_{TR}$	P_{FTR}

The mutual information, derived along the lines of information theory [7], between the random variables X and Z, given by:

$$\begin{aligned} \mathscr{I}(X;Z) &= H(X) - H(X|Z) \\ &= \sum_{x,z} \mathscr{P}\{X = x, Z = z\} \log \frac{\mathscr{P}\{X = x, Z = z\}}{\mathscr{P}\{X = x\}\mathscr{P}\{Z = z\}}, \end{aligned} \tag{5.53}$$

where $H(X)$ is the entropy of X and $H(X|Z)$ is the conditional entropy of X given Z. Using Bayes' rule, we compute the following joint probabilities:

$$\begin{aligned}\mathscr{P}\{X=T,Z=Z_1\} &= \mathscr{P}\{Z=Z_1|X=T\}\mathscr{P}\{X=T\}=pP_{TR},\\ \mathscr{P}\{X=T,Z=Z_2\} &= \mathscr{P}\{Z=Z_2|X=T\}\mathscr{P}\{X=T\}=p(1-P_{TR}),\\ \mathscr{P}\{X=NT,Z=Z_1\} &= \mathscr{P}\{Z=Z_1|X=NT\}\mathscr{P}\{X=NT\}=(1-p)(1-P_{FTR}),\\ \mathscr{P}\{X=NT,Z=Z_2\} &= \mathscr{P}\{Z=Z_2|X=NT\}\mathscr{P}\{X=NT\}=(1-p)P_{FTR}.\end{aligned} \tag{5.54}$$

Hence, the unconditional probabilities,

$$\begin{aligned}\mathscr{P}\{Z=Z_1\} &= \mathscr{P}\{X=T,Z=Z_1\}+\mathscr{P}\{X=NT,Z=Z_1\}\\ &= pP_{TR}+(1-p)(1-P_{FTR}),\\ \mathscr{P}\{Z=Z_2\} &= p(1-P_{TR})+(1-p)P_{FTR}.\end{aligned} \tag{5.55}$$

Using (5.54) and (5.55), we see that the mutual information takes the form:

$$\begin{aligned}\mathscr{I}(X;Z) = pP_{TR}\log\frac{P_{TR}}{pP_{TR}+(1-p)(1-P_{FTR})}&+\\ p(1-P_{TR})\log\frac{1-P_{TR}}{p(1-P_{TR})+(1-p)P_{FTR}}&+\\ (1-p)(1-P_{FTR})\log\frac{1-P_{FTR}}{pP_{TR}+(1-p)(1-P_{FTR})}&+\\ (1-p)P_{FTR}\log\frac{P_{FTR}}{p(1-P_{TR})+(1-p)P_{FTR}}&.\end{aligned} \tag{5.56}$$

References

1. Altman, E., Gaujal, B., Hordijk, A., Koole, G.: Optimal admission, routing and service assignment control: the case of single buffer queues. In: Proc. 37th IEEE Conf. Decision and Control, Tampa, pp. 2119–2124 (1998)
2. Axsäter, S.: State aggregation in dynamic programming: An application to scheduling of independent jobs on parallel processors. Oper. Res. Letters 2, 171–176 (1983)
3. Bean, J.C., Birge, J.R., Smith, R.L.: Aggregation in dynamic programming. Oper. Res. 35, 215–220 (1987)
4. Bellman, R.E.: Dynamic Programming. Princeton University Press, Princeton (1957)
5. Bertsimas, D., van Ryzin, G.: The dynamic traveling repairman problem. MIT Sloan School Working Paper No. 3036-89-MS (1989), `http://dspace.mit.edu/bitstream/handle/1721.1/2256/SWP-3036-20441350.pdf`
6. Browne, S., Yechiali, U.: Dynamic scheduling in single-server multiclass service systems with unit buffers. Naval Research Logistics 38, 383–396 (1991)
7. Cover, T.M., Thomas, J.A.: Elements of Information Theory, 2nd edn. Wiley-Interscience (2006)
8. Darbha, S., Krishnamoorthy, K., Pachter, M., Chandler, P.: State aggregation based linear programming approach to approximate dynamic programming. In: Proc. IEEE Conf. Decision and Control, Atlanta, GA, pp. 935–941 (2010)

9. De Farias, D.P., Van Roy, B.: The linear programming approach to approximate dynamic programming. Oper. Res., 850–865 (2003)
10. Denardo, E.V.: On linear programming in a Markov decision problem. Management Sci. 16(5), 282–288 (1970)
11. d'Epenoux, F.: A probabilistic production and inventory problem. Management Sci. 10(1), 98–108 (1963)
12. Gordon, G.: Approximate solutions to Markov decision processes. Ph.D. thesis, Carnegie Mellon University, Pittsburg, PA (1999)
13. Grötschel, M., Holland, O.: Solution of large-scale symmetric travelling salesman problems. Math. Programming 51, 141–202 (1991)
14. Grötschel, M., Lovász, L., Schijver, A.: The ellipsoid method and its consequences in combinatorial optimization. Combinatorica 1(2), 169–197 (1981)
15. Harel, A., Stulman, A.: Polling, greedy and horizon servers on a circle. Oper. Res. 43(1), 177–186 (1995)
16. Hordijk, A., Kallenberg, L.C.M.: Linear programming and Markov decision chains. Management Sci. 25(4), 352–362 (1979)
17. Howard, R.A.: Dynamic Programming and Markov Processes. The MIT Press, Cambridge (1960)
18. Kish, B., Pachter, M., Jacques, D.: Effectiveness Measures for Operations in Uncertain Environments. In: UAV Cooperative Decision and Control: Challenges and Practical Approaches, pp. 103–124. SIAM (2009)
19. Kleinrock, L.: Queueing Systems. In: Queueing Sytems. Theory, vol. I. Wiley (1975)
20. Krishnamoorthy, K., Pachter, M., Chandler, P.: Maximizing the throughput of a patrolling UAV by dynamic programming. In: Proc. IEEE Multi-Systems Conf., pp. 916–920. Denver, CO (2011)
21. Krishnamoorthy, K., Pachter, M., Chandler, P., Casbeer, D., Darbha, S.: UAV perimeter patrol operations optimization using efficient dynamic programming. In: Proc. American Control Conf., San Fransisco, CA, pp. 462–467 (2011)
22. Krishnamoorthy, K., Pachter, M., Chandler, P., Darbha, S.: Optimization of perimeter patrol operations using Unmanned Aerial Vehicles. AIAA J. Guidance, Control and Dynamics 35(2), 434–441 (2012), doi:10.2514/1.54720
23. Krishnamoorthy, K., Pachter, M., Darbha, S., Chandler, P.: Approximate dynamic programming with state aggregation applied to UAV perimeter patrol. Internat. J. Robust and Nonlinear Control 21, 1396–1409 (2011)
24. Krishnamoorthy, K., Park, M., Pachter, M., Chandler, P., Darbha, S.: Bounding procedure for stochastic dynamic programs with application to the perimeter patrol problem. In: Proc. American Control Conf., Montreal, QC, pp. 5874–5880 (2012)
25. Levy, H., Sidi, M.: Polling systems: Applications, modeling and optimization. IEEE Trans. Communications 38(10), 1750–1760 (1990)
26. Mack, C., Murphy, T., Webb, N.L.: The efficiency of N machines uni-directionally patrolled by one operative when walking time and repair times are constants. J. Royal Statistical Society, Ser. B 19(1), 166–172 (1957)
27. MacQueen, J.B.: A Modified Dynamic Programming Method for Markovian Decision Problems. J. Math. Anal. and Appl. 14, 38–43 (1966)
28. Manne, A.S.: Linear programming and sequential decisions. Management Sci. 6(3), 259–267 (1960)
29. Mendelssohn, R.: Improved bounds for aggregated linear programs. Oper. Res. 28(6), 1450–1453 (1980)
30. Mendelssohn, R.: An iterative aggregation procedure for Markov decision processes. Oper. Res. 30(1), 62–73 (1982)

31. Morrison, J.R., Kumar, P.R.: New linear program performance bounds for queueing networks. J. Optim. Theory and Appl. 100(3), 575–597 (1999)
32. Park, M., Krishnamoorthy, K., Darbha, S., Pachter, M., Chandler, P.: State aggregation based linear program for stochastic dynamic programs: an invariance property. Oper. Res. Letters (2012), doi:10.1016/j.orl.2012.08.006
33. Porteus, E.L.: Bounds and transformations for discounted finite Markov decision chains. Oper. Res. 23(4), 761–784 (1975)
34. Schuurmans, D., Patrascu, R.: Direct value-approximation for factored MDPs. In: Advances in Neural Information Processing Systems, vol. 14, pp. 1579–1586. MIT Press, Cambridge (2001)
35. Schweitzer, P.J., Seidmann, A.: Generalized polynomial approximations in Markovian decision processes. J. Math. Anal. and Appl. 110(2), 568–582 (1985)
36. Sennott, L.I.: Stochastic Dynamic Programming and the Control of Queueing Systems: Introduction. Wiley Series in Probability and Statistics. Wiley-Interscience (1999)
37. Takagi, H.: Queueing analysis of polling models. ACM Computing Surveys 20(1), 5–28 (1988)
38. Takagi, H.: Queueing analysis of polling models: progress in 1990-94. In: Frontiers in Queueing: Models and Applications in Science and Engineering, pp. 119–146. CRC Press (1997)
39. Takagi, H.: Analysis and Application of Polling Models. In: Reiser, M., Haring, G., Lindemann, C. (eds.) Performance Evaluation. LNCS, vol. 1769, pp. 423–442. Springer, Heidelberg (2000)
40. Trick, M., Zin, S.: A linear programming approach to solving stochastic dynamic programs (1993)
41. Trick, M., Zin, S.: Spline approximation to value functions: A linear programming approach. Macroeconomic Dynamics 1, 255–277 (1997)
42. Van Roy, B.: Performance loss bounds for approximate value iteration with state aggregation. Math. Oper. Res. 31(2), 234–244 (2006)

Chapter 6
A Framework for Coordination in Distributed Stochastic Systems: Perfect State Feedback and Performance Risk Aversion

Khanh Pham

Abstract. This research article considers a class of distributed stochastic systems where interconnected systems closely keep track of reference signals issued by a coordinator. Much of the existing literature concentrates on conducting decisions and control synthesis based solely on expected utilities and averaged performance. However, research in psychology and behavioral decision theory suggests that performance risk plays an important role in shaping preferences in decisions under uncertainty. Thus motivated, a new equilibrium concept, called "person-by-person equilibrium" for local best responses is proposed for analyzing signaling effects and mutual influences between an incumbent system, its coordinator and immediate neighbors. Individual member objectives are defined by the multi-attribute utility functions that capture both performance expectation and risk measures to model the satisfaction associated with local best responses with risk-averse attitudes. The problem class and approach of coordination control of distributed stochastic systems proposed here are applicable to and exemplified in military organizations and flexibly autonomous systems.

6.1 Introduction

Control and coordination of distributed stochastic systems offers a framework to analyzing intertemporal strategic interactions between individual agents or controllers, one for each interconnected systems and based on local observations. The importance of evaluating approaches in a dynamic setting and the broad flexibility and adaptability of the decision and control architectures of distributed control with communications has spurred many large-scale applications such as military command and control hierarchies, spacecraft constellations, remotely piloted platform

Khanh Pham
Air Force Research Laboratory - Space Vehicles Directorate,
3550 Aberdeen Ave. SE, Kirtland Air Force Base, NM 87117 USA
e-mail: AFRL.RVSV@kirtland.af.mil

F. Fahroo et al. (Eds.): *Recent Adv. in Res. on Unmanned Aerial Veh.*, LNCIS 444, pp. 147–179.
DOI: 10.1007/978-3-642-37694-8_6

formations, teams of humans and autonomous robots, etc. where each member can be in best response to its neighbor actions and yet has no influence on other members to which it has no communication supports.

Despite the broad interest in distributed systems, there remain significant hurdles in applying them to practical problems of interest. Interplay between common team objectives and individual member objectives can yield surprises and complex behaviors. Hence, a form of coordination control that helps balance between cooperative goals and adversarial behavior in addition of fundamentals for team and individual decisions, is necessarily required.

Thus motivated, this research article proposes a new framework and analysis to study risk-averse control of a distributed stochastic system, in particular coordination control with risk-averse attitudes toward performance uncertainty and robustness. The approach of noncooperative game-theoretic decision making and optimization is suited to coordination control, where a distributed stochastic system is distinguished into a coordinator (also known as dominant player) with significant reference signals and incumbent systems (also known as nondominant players) with fringe couplings. To account for uncertainty in inherent design problem and in preference assessment, a multi-attribute utility function that enables incumbent systems' decision makers or controllers to select the best risk-averse strategy for the attribute tradeoffs between performance expectation and risks, is therefore considered. Notice that this dominant/nondominant game structure is also prevalent in both economics [1] and social sciences [2].

The game-theoretic model of mixed player behaviors considered herein is particularly related to the research [3] that has extended the large population linear-quadratic-Gaussian games to include a major player and a large number of minor players. As such, minor players are more sensitive to variations in the behavior of major player than those of individual minor players. To overcome the curse of dimensionality, computational concerns have typically resorted the analysis to the so-called Nash certainty equivalence method, where the key idea is to break the large population game into a family of limiting two-player games. The synthesis of decentralized strategies is obtained via a set of aggregate quantities giving the mean field approximation. In contrast with such existing literature, this appealing research which is the extension of recent accounts [4] and [5] investigates: i) a stochastic dynamic game model of behavior where nondominant players not only keep track closely of the large impact by the dominant player but also monitor rivals from the peers in a less detailed way and ii) a computationally tractable model of payoff uncertainty forecast for which sufficient statistics summarize all payoff relevant information and thus are used in the person-by-person equilibrium strategies by nondominant players.

In summary, the proposed game-theoretic framework is prevalent in distributed stochastic systems with a dominant/fringe coordination structure, capturing the attributes that are important to inherent design problem and preference assessment uncertainties, their tradeoff behavior over these attributes and their risk attitude. The rest of this article is organized as follows. Section 2 introduces a new computationally tractable model for distributed stochastic systems with state-space representations of a dominant coordinator and many nondominant systems. In addition, the preliminary results on sufficient mathematical statistics that summarize all performance measure or utility relevant history and for which the person-by-person equilibrium strategies are optimal for nondominant systems are discussed in great details. Section 3 contains precise problem statements for coordination control analysis and decision optimization for the person-by-person equilibrium or feedback Nash strategy concerned by autonomous agents and incumbent systems. The construction of person-by-person strategies is established in Section 4 while some conclusions and future research directions are drawn in Section 5.

6.2 Problem Formulation

Before going into a formal presentation, it is necessary to consider some conceptual notations in this article. For instance, time t is modeled as continuous and the notation of the time interval is $[t_0, t_f]$. All random variables are defined on a probability space $(\Omega, \mathscr{F}, \mathscr{P})$ which is a triple consisting of a set Ω, a σ-algebra $\mathscr{F}$ and a probability measure $\mathscr{P} : \mathscr{F} \mapsto [0,1]$ and is equiped with a filtration $\{\mathscr{F}_t : t \in [t_0, t_f]\}$. In addition, for a given Hilbert space X with norm $||\cdot||_X$, $1 \le p \le \infty$, a Banach space is defined as follows

$$\mathscr{L}^p_{\mathscr{F}}(t_0, t_f; X) \triangleq \Big\{ \phi : [t_0, t_f] \times \Omega \mapsto X \text{ is an } X\text{-valued } \mathscr{F}_t\text{-measurable process} \\ \text{with } E\left\{ \int_{t_0}^{t_f} ||\phi(t,\omega)||^p_X dt \right\} < \infty \Big\} \tag{6.1}$$

with norm

$$||\phi(\cdot)||_{\mathscr{F},p} \triangleq \left(E\left\{ \int_{t_0}^{t_f} ||\phi(t,\omega)||^p_X dt \right\} \right)^{1/p}. \tag{6.2}$$

Furthermore, the Banach space of X-valued continuous functionals on $[t_0, t_f]$ with the max-norm induced by $||\cdot||_X$ is denoted by $\mathscr{C}(t_0, t_f; X)$. The deterministic version of (6.1) and its associated norm (6.2) is written as $\mathscr{L}^p(t_0, t_f; X)$ and $||\cdot||_p$.

A distributed stochastic system that evolves over $[t_0, t_f]$ captures interactions among a coordinator and finite number of incumbent systems. Each incumbent system that enters the distributed system is assigned a unique positive integer-valued index. The set of indices of incumbent systems is denoted by $\bar{I} \triangleq \{1, 2, \ldots, N\}$ and a typical element by i. The set of immediate neighbors associated with an incumbent system i is denoted by N_i. For concreteness, the heterogeneity of incumbent system i and $i \in \bar{I}$ is distinguished by an individual state that is governed by the stochastic differential equation with the initial-value condition $x_i(t_0) = x_i^0$

$$dx_i(t) = (A_{ii}(t)x_i(t) + B_{ii}(t)u_i(t) + C_{ii}(t)z_i(t) + \sum_{j=1}^{N_i} B_{ij}(t)u_{ij}(t))dt + G_i(t)dw_i(t) \tag{6.3}$$

where continuous-time coefficients $A_{ii} \in \mathscr{C}(t_0,t_f;\mathbb{R}^{n_i \times n_i})$, $B_{ii} \in \mathscr{C}(t_0,t_f;\mathbb{R}^{n_i \times m_i})$, $C_{ii} \in \mathscr{C}(t_0,t_f;\mathbb{R}^{n_i \times q_i})$, $B_{ij} \in \mathscr{C}(t_0,t_f;\mathbb{R}^{n_i \times r_i})$ and $G_i \in \mathscr{C}(t_0,t_f;\mathbb{R}^{n_i \times p_i})$ are deterministic matrix-valued functions. At time t, the recursive state of incumbent system i is denoted by $x_i \in \mathscr{L}^2_{\mathscr{F}_i}(t_0,t_f;\mathbb{R}^{n_i})$ with the initial state $x_i^0 \in \mathbb{R}^{n_i}$ known. The control policies from agent i to that system i are presented by $u_i \in \mathscr{L}^2_{\mathscr{F}_i}(t_0,t_f;\mathbb{R}^{m_i})$ and $z_i \in \mathscr{L}^2_{\mathscr{F}_i}(t_0,t_f;\mathbb{R}^{q_i})$. In addition, the interconnection inputs of that incumbent system i supported by the communication paths from immediate neighbors j and $j \in N_i$ are viewed as the real-valued functions $u_{ij}(t)dt$ of the following random processes

$$du_{ij}(t) = (C_{ij}(t)x_j(t) + D_{ij}(t)u_j(t))dt + dv_j(t)\,, \quad j \in N_i \tag{6.4}$$

where continuous-time coefficients $C_{ij} \in \mathscr{C}(t_0,t_f;\mathbb{R}^{r_i \times n_j})$ and $D_{ij} \in \mathscr{C}(t_0,t_f;\mathbb{R}^{r_i \times m_j})$ are deterministic matrix-valued functions. As the number of incumbent systems grows large, it is unrealistic to believe that binding agents i associated with incumbent systems i and $i \in \overline{I}$ are capable of monitoring the evolution of their immediate neighbors. Instead, it is reasonable to assume that incumbent systems only keep track of actual interactions or signaling references provided by coordinator c and $c \in \overline{I}_c$, where the set of partaking coordinators is predetermined and does not change over time.

In coordination control a coordinator c issues reference signals to two or more incumbent systems i and $i \in \overline{I}$ such that

$$z_{ic}(t)dt = (A_{ic}(t)x_c(t) + B_{ic}(t)u_c(t))dt + G_{ic}(t)dv_c(t) \tag{6.5}$$

but the incumbent systems i do not directly send signals to the coordinator c. In practice, it is further desirable to have decentralized decision making without intensive communication overheads. A potential alternative therefore involves the selection of a crude model of reduced order for the interactions among coordinator c and binding agents i associated with incumbent systems i. The actual reference signals imposed by coordinator c are now approximated by an explicit model-following of the type

$$dz_{ic}(t) = (A_{ic}(t)z_{ic}(t) + B_{ic}(t)u_c(t))dt + G_{ic}(t)dv_c(t)\,, \quad z_{ic}(t_0) = 0 \tag{6.6}$$

whereby continuous-time coefficients $A_{ic} \in \mathscr{C}(t_0,t_f;\mathbb{R}^{q_i \times q_i})$, $B_{ic} \in \mathscr{C}(t_0,t_f;\mathbb{R}^{q_i \times m_c})$ and $G_{ic} \in \mathscr{C}(t_0,t_f;\mathbb{R}^{q_i \times q_c})$ are deterministic matrix-valued function and potentially come from a structural decomposition of a monolithic distributed system with centralized dynamics.

In the state-space representation (6.3) and (6.6) one postulates independent Wiener processes $w_i(t) \triangleq w_i(t,\omega_i) : [t_0,t_f] \times \Omega_i \mapsto \mathbb{R}^{p_i}$ and $v_c(t) \triangleq v_c(t,\omega_c) : [t_0,t_f] \times \Omega_c \mapsto \mathbb{R}^{q_c}$ defined by the underlying filtered probability spaces $(\Omega_i, \mathscr{F}_i, \{\mathscr{F}_i\}_t, \mathscr{P}_i)$ and $(\Omega_c, \mathscr{F}_c, \{\mathscr{F}_c\}_t, \mathscr{P}_c)$ with the correlations of independent increments

$$E\left\{[w_i(\tau_1) - w_i(\tau_2)][w_i(\tau_1) - w_i(\tau_2)]^T\right\} = W_i|\tau_1 - \tau_2|, \quad W_i > 0, \quad \tau_1, \tau_2 \in [t_0,t_f]$$
$$E\left\{[v_c(\tau_1) - v_c(\tau_2)][v_c(\tau_1) - v_c(\tau_2)]^T\right\} = V_c|\tau_1 - \tau_2|, \quad V_c > 0, \quad \tau_1, \tau_2 \in [t_0,t_f]$$

approximate the inherent design system uncertainty due to variability and lack of knowledge.

Furthermore, the model primitives of the state recursion (6.3) in the absence of links from the immediate neighbors and environmental disturbances are also assumed to be uniformly exponentially stable. For instance, there exist positive constants η_1 and η_2 such that the pointwise matrix norm of the closed-loop state transition matrix associated with incumbent system (6.3) satisfies the inequality

$$||\Phi_i(t,\tau)|| \leq \eta_1 e^{-\eta_2(t-\tau)} \qquad \forall\, t \geq \tau \geq t_0 \ .$$

The pair $(A_{ii}(t), [B_{ii}(t), C_{ii}(t)])$ is pointwise stabilizable if there exist bounded matrix-valued functions $K_{x_i}(t)$ and $K_{z_i}(t)$ so that the closed-loop system $dx_i(t) = (A_{ii}(t) + B_{ii}(t)K_{x_i}(t) + C_{ii}(t)K_{z_i}(t))\,x_i(t)dt$ is uniformly exponentially stable.

With the local agent dynamics (6.3) considered herein, each agent i associated with incumbent system i only plays a local dynamical game with its immediate neighbors $j \in N_i$. Mutual influence controlled by the control policies from the immediate neighbors of agent i is defined by $u_{-i} \triangleq \{u_{ij} : j \in N_i\}$. Assuming its coalition N_i conveys mutual influence information u_{-i}, agent i selects, at each time instant, a tuple of control policies to optimize its multi-attribute utility function. The tuple of control laws is defined by the control processes u_i and z_i, of which z_i is supposed to follow the prediction process z_{ic} for the reference signals from coordinator c. Thus, the subsequent states of agent i is determined by its current individual states x_i and z_{ic}, its chosen action (u_i, z_i) and the coalition effects u_{-i}. In fact, the selected action (u_i, z_i) will depend on agent i's individual states x_i and z_{ic} as well as the coalition effects u_{-i}.

To further illustrate the applicability of the coordination control framework as proposed here, the classes of admissible control policies associated with (6.3) are defined by $U_i \times Z_i \subset \mathcal{L}^2_{\mathcal{F}_i}(t_0, t_f; \mathbb{R}^{m_i}) \times \mathcal{L}^2_{\mathcal{F}_{mi}}(t_0, t_f; \mathbb{R}^{q_i})$. For any given coalition effects u_{-i}, the 3-tuple $(x_i(\cdot), u_i(\cdot), z_i(\cdot))$ shall be referred to as an admissible 3-tuple if $x_i(\cdot) \in \mathcal{L}^2_{\mathcal{F}_i}(t_0, t_f; \mathbb{R}^{n_i})$ is the solution trajectory of the stochastic differential equation (6.3) when $u_i(\cdot) \in U_i$ and $z_i(\cdot) \in Z_i$.

Next, agent i evaluates its performance and makes control policies that are consistent with its preferences. There are performance tradeoffs among the closeness of local states x_i from desired states ζ_i, the size of local actions u_i and the closeness of interaction enforcements between z_i and z_{ic} on incumbent system i by coordinator c. Henceforth, agent i must carefully balance the three in order to achieve its local performance measure. Mathematically, there assumes existence of an integral-quadratic form (IQF) performance-measure $J_i : U_i \times Z_i \mapsto \mathbb{R}_+$

$$\begin{aligned} J_i(u_i, z_i; u_{-i}) &= [x_i(t_f) - \zeta_i(t_f)]^T Q_i^f [x_i(t_f) - \zeta_i(t_f)] \\ &+ \int_{t_0}^{t_f} \{x_i^T(\tau) Q_{ii}(\tau) x_i(\tau) + [x_i(\tau) - \zeta_i(\tau)]^T Q_i [x_i(\tau) - \zeta_i(\tau)]\} d\tau \\ &+ \int_{t_0}^{t_f} \{u_i^T(\tau) R_{ii}(\tau) u_i(\tau) + [z_i(\tau) - z_{ic}(\tau)]^T R_{zi}(\tau) [z_i(\tau) - z_{ic}(\tau)]\} d\tau \end{aligned} \tag{6.7}$$

where the deterministic matrix-valued functions $Q_i^f \in \mathbb{R}^{n_i \times n_i}$, $Q_{ii} \in \mathscr{C}(t_0,t_f;\mathbb{R}^{n_i \times n_i})$, $Q_i \in \mathscr{C}(t_0,t_f;\mathbb{R}^{n_i \times n_i})$ $R_{ii} \in \mathscr{C}(t_0,t_f;\mathbb{R}^{m_i \times m_i})$ and $R_{zi} \in \mathscr{C}(t_0,t_f;\mathbb{R}^{q_i \times q_i})$ representing design parameters for terminal states, transient states, control efforts and reference mismatches are positive semidefinite with $R_{ii}(t)$ and $R_{zi}(t)$ invertible.

Amongst of some research issues for coordination control which are currently under investigation is how to carry out optimal control synthesis for coordination control of distributed stochastic systems. The approach to handle the problem with a tuple of two or more control laws is to use the noncooperative game-theoretic paradigm. Particularly, an N-tuple policies $\{(u_1^*,z_1^*),(u_2^*,z_2^*),\ldots,(u_N^*,z_N^*)\}$ is said to constitute a person-by-person equilibrium solution for the coordination control problem (6.3) and performance measure (6.7) if

$$J_i^* \triangleq J_i(u_i^*,z_i^*;u_{-i}^*) \le J_i(u_i,z_i;u_{-i}^*)\,, \qquad \forall i \in \bar{I}\,. \tag{6.8}$$

That is, none of the N agents can deviate unilaterally from the equilibrium policies and gain from doing so. The justification for the restriction to such an equilibrium is that the coalition effects u_{-i}^* sent to agent i does not necessarily support its preference optimization. Therefore, they cannot do better than behave as if they strive for this equilibrium. It is reasonable to conclude that a person-by-person equilibrium of distributed control is identical to the concept of a Nash equilibrium within a noncooperative game-theoretic setting.

Because admissible feedback policy sets for agent i are not discussed, the determination of a person-by-person equilibrium for the distributed stochastic system is still not straightforward. Therefore, a further restriction is imposed next. Given the linear-quadratic properties of the state-space description (6.3) and (6.7), attention is then focused on the search for linear time-varying feedback policies generated from the locally accessible state $x_i(t)$ by

$$u_i(t) = K_{x_i}(t)x_i(t) + p_{x_i}(t) \tag{6.9}$$

$$z_i(t) = K_{z_i}(t)x_i(t) + p_{z_i}(t)\,, \qquad t \in [t_0,t_f] \tag{6.10}$$

with $K_{x_i} \in \mathscr{C}(t_0,t_f;\mathbb{R}^{m_i \times n_i})$, $K_{z_i} \in \mathscr{C}(t_0,t_f;\mathbb{R}^{q_i \times n_i})$, $p_{x_i} \in \mathscr{C}(t_0,t_f;\mathbb{R}^{m_i})$ and $p_{z_i} \in \mathscr{C}(t_0,t_f;\mathbb{R}^{q_i})$ admissible feedback policy parameters whose further defining properties will be stated shortly.

For the given (t_0,x_{ai}^0) and subject to the feedback control policies (6.9)-(6.10), agent i forms a local awareness of its state recursion (6.3) as follows

$$dx_{ai}(t) = (A_{ai}(t)x_{ai}(t) + l_{ai}(t))dt + G_{ai}(t)dw_{ai}(t)\,, \quad x_{ai}(t_0) = x_{ai}^0 \tag{6.11}$$

in which the aggregate Wiener process $w_{ai}(t) \triangleq \left[\, w_i^T(t)\;\, v_c^T(t) \,\right]^T$ has the correlations of independent increments $E\left\{[w_{ai}(\tau_1) - w_{ai}(\tau_2)][w_{ai}(\tau_1) - w_{ai}(\tau_2)]^T\right\} = W_{ai}|\tau_1 - \tau_2|$ for all $\tau_1, \tau_2 \in [t_0,t_f]$ and $W_{ai} > 0$; whereas the augmented state variable x_{ai}, its initial-valued condition x_{ai}^0, the system coefficients and parameters are defined by

$$x_{ai}(t) \triangleq \begin{bmatrix} x_i(t) \\ z_{ic}(t) \end{bmatrix}; \quad x_{ai}^0 \triangleq \begin{bmatrix} x_i^0 \\ 0 \end{bmatrix}; \quad G_{ai}(t) \triangleq \begin{bmatrix} G_i(t) & 0 \\ 0 & G_{ic}(t) \end{bmatrix}; \quad W_{ai} \triangleq \begin{bmatrix} W_i & 0 \\ 0 & V_c \end{bmatrix}$$

$$A_{ai}(t) \triangleq \begin{bmatrix} A_{ii}(t) + B_{ii}(t)K_{x_i}(t) + C_{ii}(t)K_{z_i}(t) & 0 \\ 0 & A_{ic}(t) \end{bmatrix}$$

$$l_{ai}(t) \triangleq \begin{bmatrix} B_{ii}(t)p_{x_i}(t) + C_{ii}(t)p_{z_i}(t) + \sum_{j=1}^{N_i} B_{ij}(t)u_{ij}^*(t) \\ B_{ic}(t)u_c(t) \end{bmatrix}.$$

Moreover, the sample function of the random performance measure (6.7) becomes

$$\begin{aligned} J_i(K_{x_i}, p_{x_i}; K_{z_i}, p_{z_i}) &= x_{ai}^T(t_f)Q_{ai}^f x_{ai}(t_f) + 2x_{ai}^T(t_f)S_{ai}^f + \zeta_i^T(t_f)Q_i^f \zeta_i(t_f) \\ &+ \int_{t_0}^{t_f} [x_{ai}^T(\tau)Q_{ai}(\tau)x_{ai}(\tau) + 2x_{ai}^T(\tau)S_{ai}(\tau) + \zeta_i^T(\tau)Q_i(\tau)\zeta_i(\tau) \\ &\qquad + p_{x_i}^T(\tau)R_{ii}(\tau)p_{x_i}(\tau) + p_{z_i}^T(\tau)R_{zi}(\tau)p_{z_i}(\tau)]d\tau \end{aligned} \quad (6.12)$$

whereby the corresponding weightings are given by

$$Q_{ai}^f \triangleq \begin{bmatrix} Q_i^f & 0 \\ 0 & 0 \end{bmatrix}; \quad S_{ai}^f \triangleq \begin{bmatrix} -Q_i^f \zeta_i(t_f) \\ 0 \end{bmatrix}; \quad S_{ai} \triangleq \begin{bmatrix} K_{x_i}^T R_{ii} p_{x_i} + K_{z_i}^T R_{zi} p_{z_i} - Q_i \zeta_i \\ -R_{zi} p_{z_i} \end{bmatrix}$$

$$Q_{ai} \triangleq \begin{bmatrix} Q_{ii} + Q_i + K_{x_i}^T R_{ii} K_{x_i} + K_{z_i}^T R_{zi} K_{z_i} & -2K_{z_i}^T R_{zi} \\ 0 & R_{zi} \end{bmatrix}.$$

In views of the linear-quadratic structure of the problem (6.11) and (6.12), the performance measure (6.12) is clearly a random variable with chi-squared type. Consequently, agent i adjusts its objective values into an abstract notion of satisfaction. In this research, performance expectations and risks are incorporated into a multi-attribute utility function. Then, the next task involves measuring a finite number of higher-order statistics associated with (6.12) or efficiently computing them to understand their importance. Fortunately, the research on performance assessment uncertainty offers a body of mathematical constructs that provides a starting point for such a knowledge extraction in terms of performance-measure statistics [6] and [7].

Theorem 1 *Performance-Measure Statistics.*
Let the pairs (A_{ii}, B_{ii}) and (A_{ii}, C_{ii}) be uniformly stabilizable on $[t_0, t_f]$ in the incumbent system i and $i \in \bar{I}$ governed by (6.11) and (6.12). Then for the given initial condition (t_0, x_i^0), incumbent agent i obtains the k_i-th cumulant associated with (6.12)

$$\kappa_{k_i}^i = (x_{ai}^0)^T H_i(t_0, k_i)x_{ai}^0 + 2(x_{ai}^0)^T \breve{D}_i(t_0, k_i) + D_i(t_0, k_i), \qquad k_i \in \mathbb{N} \quad (6.13)$$

whereby the supporting variables $\{H_i(s,r)\}_{r=1}^{k_i}$, $\{\breve{D}_i(s,r)\}_{r=1}^{k_i}$ and $\{D_i(s,r)\}_{r=1}^{k_i}$ satisfy the time-backward differential equations (with the dependence of $H_i(s,r)$, $\breve{D}_i(s,r)$ and $D_i(s,r)$ upon the admissible K_{x_i}, K_{z_i}, p_{x_i} and p_{z_i} suppressed)

$$\frac{d}{ds}H_i(s,1) = -A_{ai}^T(s)H_i(s,1) - H_i(s,1)A_{ai}(s) - Q_{ai}(s) \quad (6.14)$$

$$\frac{d}{ds}H_i(s,r) = -A_{ai}^T(s)H_i(s,r) - H_i(s,r)A_{ai}(s) \quad (6.15)$$

$$-\sum_{v=1}^{r-1}\frac{2r!}{v!(r-v)!}H_i(s,v)G_{ai}(s)W_{ai}G_{ai}^T(s)H_i(s,r-v), \quad 2 \le r \le k_i$$

$$\frac{d}{ds}\breve{D}_i(s,1) = -A_{ai}^T(s)\breve{D}_i(s,1) - H_i(s,1)l_{ai}(s) - S_{ai}(s) \quad (6.16)$$

$$\frac{d}{ds}\breve{D}_i(s,r) = -A_{ai}^T(s)\breve{D}_i(s,r) - H_i(s,r)l_{ai}(s), \quad 2 \le r \le k_i \quad (6.17)$$

$$\frac{d}{ds}D_i(s,1) = -\mathrm{Tr}\{H_i(s,1)G_{ai}(s)W_{ai}G_{ai}^T(s)\} - 2\breve{D}_i^T(s,1)l_{ai}(s)$$

$$- p_{x_i}^T(s)R_{ii}(s)p_{x_i}(s) - p_{z_i}^T(s)R_{zi}(s)p_{z_i}(s) - \zeta_i^T(s)Q_i(s)\zeta_i(s) \quad (6.18)$$

$$\frac{d}{ds}D_i(s,r) = -\mathrm{Tr}\{H_i(s,r)G_{ai}(s)W_{ai}G_{ai}^T(s)\} - 2\breve{D}_i^T(s,r)l_{ai}(s), 2 \le r \le k_i \quad (6.19)$$

whereby the terminal-value conditions $H_i(t_f,1) = Q_{ai}^f$, $H_i(t_f,r) = 0$ *for* $2 \le r \le k_i$*;* $\breve{D}_i(t_f,1) = S_{ai}^f$, $\breve{D}_i(t_f,r) = 0$ *for* $2 \le r \le k_i$*; and* $D_i(t_f,1) = \zeta_i^T(t_f)Q_i^f\zeta_i(t_f)$, $D_i(t_f,r) = 0$ *for* $2 \le r \le k_i$.

Proof. In general, the initial condition (t_0, x_{ai}^0) is parameterized by any arbitrary pair (s, x_{ai}^s). Then, for the given admissible affine inputs p_{x_i} and p_{z_i} in addition with admissible feedback gains K_{x_i} and K_{z_i}, the "running" performance measure is introduced as follows

$$J_i(s,x_{ai}^s) = x_{ai}^T(t_f)Q_{ai}^f x_{ai}(t_f) + 2x_{ai}^T(t_f)S_{ai}^f + \zeta_i^T(t_f)Q_i^f\zeta_i(t_f)$$

$$+ \int_s^{t_f}[x_{ai}^T(\tau)Q_{ai}(\tau)x_{ai}(\tau) + 2x_{ai}^T(\tau)S_{ai}(\tau) + \zeta_i^T(\tau)Q_i(\tau)\zeta_i(\tau)$$

$$+ p_{x_i}^T(\tau)R_{ii}(\tau)p_{x_i}(\tau) + p_{z_i}^T(\tau)R_{zi}(\tau)p_{z_i}(\tau)]d\tau. \quad (6.20)$$

The moment-generating function associated with agent i of (6.20) is defined by

$$\varphi_i(s,x_{ai}^s;\theta_i) \triangleq E\{\exp(\theta_i J_i(s,x_{ai}^s))\}, \quad (6.21)$$

for some small parameters θ_i in an open interval about 0. Thus, the cumulant-generating function immediately follows

$$\psi_i(s,x_{ai}^s;\theta_i) \triangleq \ln\{\varphi_i(s,x_{ai}^s;\theta_i)\}, \quad (6.22)$$

for some θ_i in some (possibly smaller) open interval about 0 while $\ln\{\cdot\}$ denotes the natural logarithmic transformation.

For notational simplicity, it is convenient to define $\varpi_i(s,x_{ai}^s;\theta_i) \triangleq \exp\{\theta_i J_i(s,x_{ai}^s)\}$ and $\varphi_i(s,x_{ai}^s;\theta_i) \triangleq E\{\varpi_i(s,x_{ai}^s;\theta_i)\}$ together with the time derivative of

$$\frac{d}{ds}\varphi_i\left(s,x_{ai}^s;\theta_i\right)=-\theta_i\Big\{(x_{ai}^s)^T Q_{ai}(s)x_{ai}^s+2(x_{ai}^s)^T S_{ai}(s)$$
$$+\zeta_i^T(s)Q_i(s)\zeta_i(s)+p_{x_i}^T(s)R_{ii}(s)p_{x_i}(s)+p_{z_i}^T(s)R_{zi}(s)p_{z_i}(s)\Big\}\varphi_i\left(s,x_{ai}^s;\theta_i\right). \quad (6.23)$$

Using the standard Ito's formula, it yields

$$\begin{aligned}d\varphi_i\left(s,x_{ai}^s;\theta_i\right)&=E\left\{d\varpi_i\left(s,x_{ai}^s;\theta_i\right)\right\},\\&=\varphi_{i,s}\left(s,x_{ai}^s;\theta_i\right)ds+\varphi_{i,x_{ai}^s}\left(s,x_{ai}^s;\theta_i\right)\left[A_{ai}(s)x_{ai}^s+l_{ai}(s)\right]ds\\&\quad+\frac{1}{2}\mathrm{Tr}\{\varphi_{i,x_{ai}^s x_{ai}^s}\left(s,x_{ai}^s;\theta_i\right)G_{ai}(s)W_{ai}G_{ai}^T(s)\}ds.\end{aligned}$$

Furthermore, the moment-generating function of (6.20) can also be expressed by

$$\varphi_i\left(s,x_{ai}^s;\theta_i\right)\triangleq\rho_i\left(s;\theta_i\right)\exp\left\{(x_{ai}^s)^T\Upsilon_i(s;\theta_i)x_{ai}^s+2(x_{ai}^s)^T\eta_i(s;\theta_i)\right\} \quad (6.24)$$

whereby all the supporting entities are going to be determined in the sequel. In particular, the partial derivatives of (6.24) results in

$$\begin{aligned}\frac{d}{ds}\varphi_i\left(s,x_{ai}^s;\theta_i\right)=&\Big\{\frac{\frac{d}{ds}\rho_i(s;\theta_i)}{\rho_i(s,\theta_i)}+(x_{ai}^s)^T\frac{d}{ds}\Upsilon_i(s;\theta_i)x_{ai}^s+2(x_{ai}^s)^T\frac{d}{ds}\eta_i(s;\theta_i)\\&+(x_{ai}^s)^T A_{ai}^T(s)\Upsilon_i(s;\theta_i)x_{ai}^s+(x_{ai}^s)^T\Upsilon_i(s;\theta_i)A_{ai}(s)x_{ai}^s+2(x_{ai}^s)^T A_{ai}^T(s)\eta_i(s;\theta_i)\\&+2(x_{ai}^s)^T\Upsilon_i(s;\theta_i)l_{ai}(s)+2\eta_i^T(s;\theta_i)l_{ai}(s)+\mathrm{Tr}\{\Upsilon_i(s;\theta_i)G_{ai}(s)W_{ai}G_{ai}^T(s)\}\\&+2(x_{ai}^s)^T\Upsilon_i(s;\theta_i)G_{ai}(s)W_{ai}G_{ai}^T(s)\Upsilon_i(s;\theta_i)x_{ai}^s\Big\}\varphi_i\left(s,x_{ai}^s;\theta_i\right). \quad (6.25)\end{aligned}$$

Equating the expression (6.23) with that of (6.25) and having both linear and quadratic terms independent of x_{ai}^s yield the following results, wherein $\upsilon_i(s;\theta_i)\triangleq\ln\{\rho_i(s;\theta_i)\}$

$$\begin{aligned}\frac{d}{ds}\Upsilon_i(s;\theta_i)=&-A_{ai}^T(s)\Upsilon_i(s;\theta_i)-\Upsilon_i(s;\theta_i)A_{ai}(s)\\&-2\Upsilon_i(s;\theta_i)G_{ai}(s)W_{ai}G_{ai}^T(s)\Upsilon_i(s;\theta_i)-\theta_iQ_{ai}(s) \quad (6.26)\\\frac{d}{ds}\eta_i\left(s;\theta_i\right)=&-A_{ai}^T(s)\eta_i(s;\theta_i)-\Upsilon_i(s;\theta_i)l_{ai}(s)-\theta_iS_{ai}(s) \quad (6.27)\\\frac{d}{ds}\upsilon_i\left(s;\theta_i\right)=&-\mathrm{Tr}\left\{\Upsilon_i(s;\theta_i)G_{ai}(s)W_{ai}G_{ai}^T(s)\right\}-2\eta_i^T(s;\theta_i)l_{ai}(s)-\theta_i\zeta_i^T(s)Q_i(s)\zeta_i(s)\\&-\theta_ip_{x_i}^T(s)R_{ii}(s)p_{x_i}(s)-\theta_ip_{z_i}^T(s)R_{zi}(s)p_{z_i}(s) \quad (6.28)\end{aligned}$$

At the final time $s=t_f$, it follows that

$$\begin{aligned}\varphi_i(t_f,x_{ai}(t_f);\theta_i)&=\rho_i(t_f;\theta_i)\exp\left\{x_{ai}^T(t_f)\Upsilon_i(t_f;\theta_i)x_{ai}(t_f)+2x_{ai}^T(t_f)\eta_i(t_f;\theta_i)\right\}\\&=E\left\{\exp\left\{\theta_i[x_{ai}^T(t_f)Q_{ai}^f x_{ai}(t_f)+2x_{ai}^T(t_f)S_{ai}^f+\zeta_i^T(t_f)Q_i^f\zeta_i(t_f)]\right\}\right\}\end{aligned}$$

which in turn yields the terminal-value conditions as $\Upsilon_i(t_f;\theta_i)=\theta_iQ_{ai}^f$; $\eta_i(t_f;\theta_i)=\theta_iS_{ai}^f$; and $\upsilon_i(t_f;\theta_i)=\theta_i\zeta_i^T(t_f)Q_i^f\zeta_i(t_f)$.

As it turns out, all the higher-order characteristic distributions associated with performance uncertainty are very well captured in higher-order performance-measure statistics associated with the chi-squared random performance measure (6.20). In views of the expression (6.24) and the definition of (6.22), the cumulant-generating function or second-order characteristic function of (6.20) is rewritten as follows

$$\psi_i(s, x_{ai}^s; \theta_i) = (x_{ai}^s)^T \Upsilon_i(s;\theta_i) x_{ai}^s + 2(x_{ai}^s)^T \eta_i(s;\theta_i) + \upsilon_i(s;\theta_i). \tag{6.29}$$

Subsequently, higher-order statistics of the random performance measure (6.20) that depict the performance uncertainty can now be determined by a Maclaurin series expansion of the cumulant-generating function (6.29); e.g.,

$$\psi_i(s, x_{ai}^s; \theta_i) = \sum_{r=1}^{\infty} \frac{\partial^{(r)}}{\partial \theta_i^{(r)}} \psi_i(s, x_{ai}^s; \theta_i)\Big|_{\theta_i=0} \frac{\theta_i^r}{r!}, \tag{6.30}$$

from which all $\kappa_r \triangleq \frac{\partial^{(r)}}{\partial \theta_i^{(r)}} \psi_i(s, x_{ai}^s; \theta_i)\Big|_{\theta_i=0}$ are known as the mathematical statistics of the chi-squared random performance measure (6.20). Moreover, the series expansion coefficients are computed by using the cumulant-generating function (6.29)

$$\begin{aligned}\frac{\partial^{(r)}}{\partial \theta_i^{(r)}} \psi_i(s, x_{ai}^s; \theta_i)\Big|_{\theta_i=0} &= (x_{ai}^s)^T \frac{\partial^{(r)}}{\partial \theta_i^{(r)}} \Upsilon_i(s;\theta_i)\Big|_{\theta_i=0} x_{ai}^s \\ &+ 2(x_{ai}^s)^T \frac{\partial^{(r)}}{\partial \theta_i^{(r)}} \eta_i(s;\theta_i)\Big|_{\theta_i=0} + \frac{\partial^{(r)}}{\partial \theta_i^{(r)}} \upsilon_i(s;\theta_i)\Big|_{\theta_i=0}.\end{aligned} \tag{6.31}$$

In view of the definition (6.30), the rth performance-measure statistic is given by

$$\begin{aligned}\kappa_r = (x_{ai}^s)^T & \frac{\partial^{(r)}}{\partial \theta_i^{(r)}} \Upsilon_i(s;\theta_i)\Big|_{\theta_i=0} x_{ai}^s \\ &+ 2(x_{ai}^s)^T \frac{\partial^{(r)}}{\partial \theta_i^{(r)}} \eta_i(s;\theta_i)\Big|_{\theta_i=0} + \frac{\partial^{(r)}}{\partial \theta_i^{(r)}} \upsilon_i(s;\theta_i)\Big|_{\theta_i=0}\end{aligned} \tag{6.32}$$

for any finite $1 \le r < \infty$. For notational convenience, the change of notations

$$H_i(s,r) \triangleq \frac{\partial^{(r)} \Upsilon_i(s;\theta_i)}{\partial \theta_i^{(r)}}\Big|_{\theta_i=0}; \breve{D}_i(s,r) \triangleq \frac{\partial^{(r)} \eta_i(s;\theta_i)}{\partial \theta_i^{(r)}}\Big|_{\theta_i=0}; D_i(s,r) \triangleq \frac{\partial^{(r)} \upsilon_i(s;\theta_i)}{\partial \theta_i^{(r)}}\Big|_{\theta_i=0}$$

is introduced. What remains is to show that the solutions $H_i(s,r)$, $\breve{D}_i(s,r)$ and $D_i(s,r)$ for $1 \le r \le k_i$ and $k_i \in \mathbb{N}$ indeed satisfy the time-backward matrix, vector and scalar-valued differential equations (6.14)-(6.19). Notice that these differential equations

(6.14)-(6.19) are readily obtained by successively taking derivatives with respect to θ_i of the cumulant-supporting equations (6.26)-(6.28) under the assumption of (A_{ii}, B_{ii}) and (A_{ii}, C_{ii}) uniformly stabilizable on the interval $[t_0, t_f]$. □

Furthermore, some attractive properties of the solutions to the cumulant-generating equations (6.14)-(6.19), for which the problem of coordination control with risk-averse performance of the class of distributed stochastic systems considered here is therefore well-posed, are presented as follows.

Theorem 2 *Existence of Solutions for Performance-Measure Statistics.*
Let the pairs $(A_{ii}(\cdot), B_{ii}(\cdot))$ and $(A_{ii}(\cdot), C_{ii}(\cdot))$ be uniformly stabilizable. Then, for any given $k_i \in \mathbb{N}$, the cumulant-generating equations (6.14)-(6.19) admit unique and bounded solutions $\{H_i(\cdot,r)\}_{r=1}^{k_i}$, $\{\breve{D}_i(\cdot,r)\}_{r=1}^{k_i}$ and $\{D_i(\cdot,r)\}_{r=1}^{k_i}$ on $[t_0, t_f]$.

Proof. Under the assumption of stabilizability, there always exist some feedback parameters $K_{x_i}(\cdot)$ and $K_{z_i}(\cdot)$ such that the continuous-time aggregate state matrix $A_{ai}(\cdot)$ is exponentially stable on $[t_0, t_f]$. According to the results in [8], the state transition matrix $\Phi_{ai}(t, t_0)$, associated with the continuous-time composite state matrix $A_{ai}(\cdot)$, has the following properties

$$\frac{d}{dt}\Phi_{ai}(t,t_0) = A_{ai}(t)\Phi_{ai}(t,t_0), \qquad \Phi_{ai}(t_0,t_0) = I,$$

$$\lim_{t_f \to \infty} ||\Phi_{ai}(t_f,\tau)|| = 0, \qquad \lim_{t_f \to \infty} \int_{t_0}^{t_f} ||\Phi_{ai}(t_f,\tau)||^2 d\tau < \infty.$$

By the matrix variation of constant formula, the unique solutions to the time-backward matrix differential equations (6.14)-(6.19) together with the terminal-value conditions are then written as follows

$$H_i(s,1) = \Phi_{ai}^T(t_f,s)Q_{ai}^f \Phi_{ai}(t_f,s) + \int_s^{t_f} \Phi_{ai}^T(\tau,s)Q_{ai}(\tau)\Phi_{ai}(\tau,s)d\tau$$

$$H_i(s,r) = \int_s^{t_f} \Phi_{ai}^T(\tau,s) \sum_{v=1}^{r-1} \frac{2r!}{v!(r-v)!} H_i(\tau,v)G_{ai}(\tau)W_{ai}G_{ai}^T(\tau)H_i(\tau,r-v)\Phi_{ai}(\tau,s)d\tau$$

$$\breve{D}_i(s,1) = -\Phi_{ai}^T(t_f,s)Q_i^f \zeta_i(t_f) + \int_s^{t_f} \Phi_{ai}^T(\tau,s)\{H_i(\tau,1)l_{ai}(\tau) + S_{ai}(\tau)\}d\tau$$

$$\breve{D}_i(s,r) = \int_s^{t_f} \Phi_{ai}^T(\tau,s)H_i(\tau,r)l_{ai}(\tau)d\tau, \quad 2 \le r \le k_i$$

$$D_i(s,1) = \zeta_i^T(t_f)Q_i^f \zeta_i(t_f) + \int_s^{t_f} \{\mathrm{Tr}\{H_i(\tau,1)G_{ai}(\tau)W_{ai}G_{ai}^T(\tau)\} + 2\breve{D}_i^T(\tau,1)l_{ai}(\tau)$$
$$+ p_{x_i}^T(\tau)R_{ii}(\tau)p_{x_i}(\tau) + p_{z_i}^T(\tau)R_{zi}(\tau)p_{z_i}(\tau) + \zeta_i^T(\tau)Q_i(\tau)\zeta_i(\tau)\}d\tau$$

$$D_i(s,r) = \int_s^{t_f} \{\mathrm{Tr}\{H_i(\tau,r)G_{ai}(\tau)W_{ai}G_{ai}^T(\tau)\} + 2\breve{D}_i^T(\tau,r)l_{ai}(\tau)\}d\tau, \quad 2 \le r \le k_i.$$

As long as the growth rates of the integrals are not faster than those of exponentially decreasing $\Phi_{ai}(\cdot,\cdot)$ and $\Phi_{ai}^T(\cdot,\cdot)$ factors, it is therefore concluded that there exist upper bounds on the non-negative and monotically increasing solutions $H_i(\cdot,r)$, $\breve{D}_i(\cdot,r)$ and $D_i(\cdot,r)$ for any time interval $[t_0,t_f]$.

6.3 Problem Statements

The problem of adapting to performance uncertainty is now addressed by leveraging increased insight into the roles played by performance-measure statistics (6.13). It is interesting to note that all the performance-measure statistics (6.13) are functions of time-backward evolutions and do not depend on intermediate recursive state values $x_{ai}(t)$ governed by the state-space representation (6.11)-(6.12) for incumbent agent i at each point of time $t \in [t_0,t_f]$. Henceforth, these time-backward evolutions (6.14)-(6.19) of which the admissible decision variables K_{x_i}, K_{z_i}, p_{x_i} and p_{z_i} from the 2-tuple person-by-person equilibrium strategy (6.9)-(6.10) are embedded, are therefore considered as the new dynamical equations with the associated state variables $H_i(\cdot,r)$, $\breve{D}_i(\cdot,r)$ and $D_i(\cdot,r)$, not the traditional system states $x_{ai}(\cdot)$.

To properly develop the problem statements within the concept of the person-by-person equilibrium strategy for agent i and $i \in \bar{I}$, the new dynamics (6.14)-(6.19) based upon the performance-measure statistics of (6.13) is rewritten in accordance of the following matrix partitions, for $1 \le r \le k_i$ and $k_i \in \mathbb{N}$

$$H_i(\cdot,r) \triangleq \begin{bmatrix} H_{i,r}^{11}(\cdot) & H_{i,r}^{12}(\cdot) \\ H_{i,r}^{21}(\cdot) & H_{i,r}^{22}(\cdot) \end{bmatrix}, \qquad \breve{D}_i(\cdot,r) \triangleq \begin{bmatrix} \breve{D}_{i,r}^{11}(\cdot) \\ \breve{D}_{i,r}^{21}(\cdot) \end{bmatrix}.$$

For notational simplicity, it is now useful to denote the right members of the dynamics (6.14)-(6.19) as the mappings

$$\begin{aligned}
\mathscr{F}_i^r &: [t_0,t_f] \times (\mathbb{R}^{n_i \times n_i})^{4k_i} \times \mathbb{R}^{m_i \times n_i} \times \mathbb{R}^{q_i \times n_i} \mapsto \mathbb{R}^{n_i \times n_i} \\
\mathscr{F}_i^{k_i+r} &: [t_0,t_f] \times (\mathbb{R}^{n_i \times n_i})^{4k_i} \times \mathbb{R}^{m_i \times n_i} \times \mathbb{R}^{q_i \times n_i} \mapsto \mathbb{R}^{n_i \times n_i} \\
\mathscr{F}_i^{2k_i+r} &: [t_0,t_f] \times (\mathbb{R}^{n_i \times n_i})^{4k_i} \times \mathbb{R}^{m_i \times n_i} \times \mathbb{R}^{q_i \times n_i} \mapsto \mathbb{R}^{n_i \times n_i} \\
\mathscr{F}_i^{3k_i+r} &: [t_0,t_f] \times (\mathbb{R}^{n_i \times n_i})^{4k_i} \mapsto \mathbb{R}^{n_i \times n_i} \\
\breve{\mathscr{G}}_i^r &: [t_0,t_f] \times (\mathbb{R}^{n_i \times n_i})^{4k_i} \times (\mathbb{R}^{n_i})^{k_i} \times \mathbb{R}^{m_i \times n_i} \times \mathbb{R}^{q_i \times n_i} \times \mathbb{R}^{m_i} \times \mathbb{R}^{q_i} \mapsto \mathbb{R}^{n_i} \\
\breve{\mathscr{G}}_i^{k_i+r} &: [t_0,t_f] \times (\mathbb{R}^{n_i \times n_i})^{4k_i} \times (\mathbb{R}^{n_i})^{k_i} \times \mathbb{R}^{m_i} \times \mathbb{R}^{q_i} \mapsto \mathbb{R}^{n_i} \\
\mathscr{G}_i^r &: [t_0,t_f] \times (\mathbb{R}^{n_i \times n_i})^{4k_i} \times (\mathbb{R}^{n_i})^{2k_i} \times \mathbb{R}^{m_i} \times \mathbb{R}^{q_i} \mapsto \mathbb{R}
\end{aligned}$$

with the rules of action

$$\begin{aligned}
\mathscr{F}_i^1(s,\mathscr{H}_i,K_{x_i},K_{z_i}) \triangleq & -[A_{ii}(s)+B_{ii}(s)K_{x_i}(s)+C_{ii}(s)K_{z_i}(s)]^T \mathscr{H}_i^1(s) \\
& - \mathscr{H}_i^1(s)[A_{ii}(s)+B_{ii}(s)K_{x_i}(s)+C_{ii}(s)K_{z_i}(s)] \\
& - Q_{ii}(s) - Q_i(s) - K_{x_i}^T(s)R_{ii}(s)K_{x_i}(s) - K_{z_i}^T(s)R_{zi}(s)K_{z_i}(s)
\end{aligned}$$

$$\begin{aligned}\mathscr{F}_i^r(s,\mathscr{H}_i,K_{x_i},K_{z_i}) \triangleq & -[A_{ii}(s)+B_{ii}(s)K_{x_i}(s)+C_{ii}(s)K_{z_i}(s)]^T\mathscr{H}_i^r(s)\\ & -\mathscr{H}_i^r(s)[A_{ii}(s)+B_{ii}(s)K_{x_i}(s)+C_{ii}(s)K_{z_i}(s)]\\ & -\sum_{v=1}^{r-1}\frac{2r!}{v!(r-v)!}\mathscr{H}_i^v(s)G_i(s)W_iG_i^T(s)\mathscr{H}_i^{r-v}(s)\\ & -\sum_{v=1}^{r-1}\frac{2r!}{v!(r-v)!}\mathscr{H}_i^{k_i+v}(s)G_{ic}(s)V_cG_{ic}^T(s)\mathscr{H}_i^{2k_i+r-v}(s)\end{aligned}$$

$$\begin{aligned}\mathscr{F}_i^{k_i+1}(s,\mathscr{H}_i,K_{x_i},K_{z_i}) \triangleq & -[A_{ii}(s)+B_{ii}(s)K_{x_i}(s)+C_{ii}(s)K_{z_i}(s)]^T\mathscr{H}_i^{k_i+1}(s)\\ & -\mathscr{H}_i^{k_i+1}(s)A_{ic}(s)+2K_{z_i}^T(s)R_{zi}(s)\end{aligned}$$

$$\begin{aligned}\mathscr{F}_i^{k_i+r}(s,\mathscr{H}_i,K_{x_i},K_{z_i}) \triangleq & -[A_{ii}(s)+B_{ii}(s)K_{x_i}(s)+C_{ii}(s)K_{z_i}(s)]^T\mathscr{H}_i^{k_i+r}(s)\\ & -\mathscr{H}_i^{k_i+r}(s)A_{ic}(s)-\sum_{v=1}^{r-1}\frac{2r!}{v!(r-v)!}\mathscr{H}_i^v(s)G_i(s)W_iG_i^T(s)\mathscr{H}_i^{k_i+r-v}(s)\\ & -\sum_{v=1}^{r-1}\frac{2r!}{v!(r-v)!}\mathscr{H}_i^{k_i+v}(s)G_{ic}(s)V_cG_{ic}^T(s)\mathscr{H}_i^{3k_i+r-v}(s)\end{aligned}$$

$$\begin{aligned}\mathscr{F}_i^{2k_i+1}(s,\mathscr{H}_i,K_{x_i},K_{z_i}) \triangleq & -A_{ic}^T(s)\mathscr{H}_i^{2k_i+1}(s)\\ & -\mathscr{H}_i^{2k_i+1}(s)[A_{ii}(s)+B_{ii}(s)K_{x_i}(s)+C_{ii}(s)K_{z_i}(s)]\end{aligned}$$

$$\begin{aligned}\mathscr{F}_i^{2k_i+r}(s,\mathscr{H}_i,K_{x_i},K_{z_i}) \triangleq & -A_{ic}^T(s)\mathscr{H}_i^{2k_i+r}(s)\\ & -\mathscr{H}_i^{2k_i+r}(s)[A_{ii}(s)+B_{ii}(s)K_{x_i}(s)+C_{ii}(s)K_{z_i}(s)]\\ & -\sum_{v=1}^{r-1}\frac{2r!}{v!(r-v)!}\mathscr{H}_i^{2k_i+v}(s)G_i(s)W_iG_i^T(s)\mathscr{H}_i^{r-v}(s)\\ & -\sum_{v=1}^{r-1}\frac{2r!}{v!(r-v)!}\mathscr{H}_i^{3k_i+v}(s)G_{ic}(s)V_cG_{ic}^T(s)\mathscr{H}_i^{2k_i+r-v}(s)]\end{aligned}$$

$$\mathscr{F}_i^{3k_i+1}(s,\mathscr{H}_i) \triangleq -A_{ic}^T(s)\mathscr{H}_i^{3k_i+1}(s)-\mathscr{H}_i^{3k_i+1}(s)A_{ic}(s)-R_{zi}(s)$$

$$\begin{aligned}\mathscr{F}_i^{3k_i+r}(s,\mathscr{H}_i) \triangleq & -A_{ic}^T(s)\mathscr{H}_i^{3k_i+r}(s)-\mathscr{H}_i^{3k_i+r}(s)A_{ic}(s)-R_{zi}(s)\\ & -\sum_{v=1}^{r-1}\frac{2r!}{v!(r-v)!}\mathscr{H}_i^{2k_i+v}(s)G_i(s)W_iG_i^T(s)\mathscr{H}_i^{k_i+r-v}(s)\\ & -\sum_{v=1}^{r-1}\frac{2r!}{v!(r-v)!}\mathscr{H}_i^{3k_i+v}(s)G_{ic}(s)V_cG_{ic}^T(s)\mathscr{H}_i^{3k_i+r-v}(s)\end{aligned}$$

$$\breve{\mathscr{G}}_i^1(s,\mathscr{H}_i,\breve{\mathscr{D}}_i,K_{x_i},K_{z_i},p_{x_i},p_{z_i}) \triangleq -[A_{ii}(s)+B_{ii}(s)K_{x_i}(s)+C_{ii}(s)K_{z_i}(s)]^T\breve{\mathscr{D}}_i^1(s)$$
$$-\mathscr{H}_i^1(s)[B_{ii}(s)p_{x_i}(s)+C_{ii}(s)p_{z_i}(s)+\sum_{j=1}^{N_i}B_{ij}(s)u_{ij}^*(s)]$$
$$-\mathscr{H}_i^{k_i+1}(s)B_{ic}(s)u_{ic}(s)-K_{x_i}^T(s)R_{ii}(s)p_{x_i}(s)-K_{z_i}^T R_{zi}(s)p_{z_i}(s)+Q_i(s)\zeta_i(s)$$

$$\breve{\mathscr{G}}_i^r(s,\mathscr{H}_i,\breve{\mathscr{D}}_i,K_{x_i},K_{z_i},p_{x_i},p_{z_i}) \triangleq -[A_{ii}(s)+B_{ii}(s)K_{x_i}(s)+C_{ii}(s)K_{z_i}(s)]^T\breve{\mathscr{D}}_i^r(s)$$
$$-\mathscr{H}_i^r(s)[B_{ii}(s)p_{x_i}(s)+C_{ii}(s)p_{z_i}(s)+\sum_{j=1}^{N_i}B_{ij}(s)u_{ij}^*(s)]-\mathscr{H}_i^{k_i+r}(s)B_{ic}(s)u_{ic}(s)$$

$$\breve{\mathscr{G}}_i^{k_i+1}(s,\mathscr{H}_i,\breve{\mathscr{D}}_i,p_{x_i},p_{z_i}) \triangleq -A_{ic}^T(s)\breve{\mathscr{D}}_i^{k_i+1}(s)-\mathscr{H}_i^{3k_i+1}(s)B_{ic}(s)u_{ic}(s)+R_{zi}(s)p_{z_i}(s)$$
$$-\mathscr{H}_i^{2k_i+1}(s)[B_{ii}(s)p_{x_i}(s)+C_{ii}(s)p_{z_i}(s)+\sum_{j=1}^{N_i}B_{ij}(s)u_{ij}^*(s)]$$

$$\breve{\mathscr{G}}_i^{k_i+r}(s,\mathscr{H}_i,\breve{\mathscr{D}}_i,p_{x_i},p_{z_i}) \triangleq -A_{ic}^T(s)\breve{\mathscr{D}}_i^{k_i+r}(s)-\mathscr{H}_i^{3k_i+r}(s)B_{ic}(s)u_{ic}(s)$$
$$-\mathscr{H}_i^{2k_i+r}(s)[B_{ii}(s)p_{x_i}(s)+C_{ii}(s)p_{z_i}(s)+\sum_{j=1}^{N_i}B_{ij}(s)u_{ij}^*(s)]$$

$$\mathscr{G}_i^1(s,\mathscr{H}_i,\breve{\mathscr{D}}_i,p_{x_i},p_{z_i}) \triangleq -\mathrm{Tr}\{\mathscr{H}_i^1(s)G_i(s)W_iG_i^T(s)+\mathscr{H}_i^{3k_i+1}(s)G_{ic}(s)V_cG_{ic}^T(s)\}$$
$$-2(\breve{\mathscr{D}}_i^1)^T(s)[B_{ii}(s)p_{x_i}(s)+C_{ii}(s)p_{z_i}(s)+\sum_{j=1}^{N_i}B_{ij}(s)u_{ij}^*(s)]-2(\breve{\mathscr{D}}_i^{k_i+1})^T(s)B_{ic}(s)u_{ic}(s)$$
$$-\zeta_i^T(s)Q_i(s)\zeta_i(s)-p_{x_i}^T(s)R_{ii}(s)p_{x_i}(s)-p_{z_i}^T(s)R_{zi}(s)p_{z_i}(s)$$

$$\mathscr{G}_i^r(s,\mathscr{H}_i,\breve{\mathscr{D}}_i,p_{x_i},p_{z_i}) \triangleq -\mathrm{Tr}\{\mathscr{H}_i^r(s)G_i(s)W_iG_i^T(s)+\mathscr{H}_i^{3k_i+r}(s)G_{ic}(s)V_cG_{ic}^T(s)\}$$
$$-2(\breve{\mathscr{D}}_i^r)^T(s)[B_{ii}(s)p_{x_i}(s)+C_{ii}(s)p_{z_i}(s)+\sum_{j=1}^{N_i}B_{ij}(s)u_{ij}^*(s)]-2(\breve{\mathscr{D}}_i^{k_i+r})^T(s)B_{ic}(s)u_{ic}(s)$$

whereby the components of $4k_i$-tuple $\mathscr{H}_i$, $2k_i$-tuple $\breve{\mathscr{D}}_i$ and k_i-tuple $\mathscr{D}_i$ variables are defined by

$$\begin{aligned}
\mathscr{H}_i &\triangleq (\mathscr{H}_i^1,\ldots,\mathscr{H}_i^{k_i},\mathscr{H}_i^{k_i+1},\ldots,\mathscr{H}_i^{2k_i},\mathscr{H}_i^{2k_i+1},\ldots,\mathscr{H}_i^{3k_i},\mathscr{H}_i^{3k_i+1},\ldots,\mathscr{H}_i^{4k_i})\\
&\equiv (H_{i,1}^{11},\ldots,H_{i,k_i}^{11},H_{i,1}^{12},\ldots,H_{i,k_i}^{12},H_{i,1}^{21},\ldots,H_{i,k_i}^{21},H_{i,1}^{22},\ldots,H_{i,k_i}^{22}),\\
\breve{\mathscr{D}}_i &\triangleq (\breve{\mathscr{D}}_i^1,\ldots,\breve{\mathscr{D}}_i^{k_i},\breve{\mathscr{D}}_i^{k_i+1},\ldots,\breve{\mathscr{D}}_i^{2k_i}) \equiv (\breve{D}_{i,1}^{11},\ldots,\breve{D}_{i,k_i}^{11},\breve{D}_{i,1}^{21},\ldots,\breve{D}_{i,k_i}^{21})\\
\mathscr{D}_i &\triangleq (\mathscr{D}_i^1,\ldots,\mathscr{D}_i^{k_i}) \equiv (D_i^1,\ldots,D_i^{k_i}).
\end{aligned}$$

Hence, the product system of dynamical equations, whose mapping are

$\mathscr{F}_i^1 \times \cdots \times \mathscr{F}_i^{4k_i} : [t_0, t_f] \times (\mathbb{R}^{n_i \times n_i})^{4k_i} \times \mathbb{R}^{m_i \times n_i} \times \mathbb{R}^{q_i \times n_i} \mapsto (\mathbb{R}^{n_i \times n_i})^{4k_i}$

$\breve{\mathscr{G}}_i^1 \times \cdots \times \breve{\mathscr{G}}_i^{2k_i} : [t_0, t_f] \times (\mathbb{R}^{n_i \times n_i})^{4k_i} \times (\mathbb{R}^{n_i})^{k_i} \times \mathbb{R}^{m_i \times n_i} \times \mathbb{R}^{q_i \times n_i} \times \mathbb{R}^{m_i} \times \mathbb{R}^{q_i} \mapsto (\mathbb{R}^{n_i})^{2k_i}$

$\mathscr{G}_i^1 \times \cdots \times \mathscr{G}_i^{k_i} : [t_0, t_f] \times (\mathbb{R}^{n_i \times n_i})^{4k_i} \times (\mathbb{R}^{n_i})^{2k_i} \times \mathbb{R}^{m_i} \times \mathbb{R}^{q_i} \mapsto \mathbb{R}^{k_i}$

in coordination control of the problem class with performance risk aversion, becomes

$$\frac{d}{ds}\mathscr{H}_i(s) = \mathscr{F}_i(s, \mathscr{H}_i(s), K_{x_i}(s), K_{z_i}(s)), \qquad \mathscr{H}_i(t_f) = \mathscr{H}_i^f, \tag{6.33}$$

$$\frac{d}{ds}\breve{\mathscr{D}}_i(s) = \breve{\mathscr{G}}_i(s, \mathscr{H}_i(s), \breve{\mathscr{D}}_i(s), K_{x_i}(s), K_{z_i}(s), p_{x_i}(s), p_{z_i}(s)), \quad \breve{\mathscr{D}}_i(t_f) = \breve{\mathscr{D}}_i^f, \tag{6.34}$$

$$\frac{d}{ds}\breve{\mathscr{D}}_i(s) = \breve{\mathscr{G}}_i(s, \mathscr{H}_i(s), \breve{\mathscr{D}}_i(s), p_{x_i}(s), p_{z_i}(s)), \quad \breve{\mathscr{D}}_i(t_f) = \breve{\mathscr{D}}_i^f, \tag{6.35}$$

whereby $\mathscr{F}_i \triangleq \mathscr{F}_i^1 \times \cdots \times \mathscr{F}_i^{4k_i}$, $\breve{\mathscr{G}}_i \triangleq \breve{\mathscr{G}}_i^1 \times \cdots \times \breve{\mathscr{G}}_i^{2k_i}$ and $\mathscr{G}_i \triangleq \mathscr{G}_i^1 \times \cdots \times \mathscr{G}_i^{k_i}$, in addition with the product system of the terminal-value conditions

$$\mathscr{H}_i^f \triangleq Q_i^f \times \underbrace{0 \times \cdots \times 0}_{(k_i-1)\text{-times}} \times \underbrace{0 \times \cdots \times 0}_{k_i\text{-times}} \times \underbrace{0 \times \cdots \times 0}_{k_i\text{-times}} \times \underbrace{0 \times \cdots \times 0}_{k_i\text{-times}}$$

$$\breve{\mathscr{D}}_i^f \triangleq -Q_i^f \zeta_i(t_f) \times \underbrace{0 \times \cdots \times 0}_{(k_i-1)\text{-times}} \times \underbrace{0 \times \cdots \times 0}_{k_i\text{-times}}$$

$$\mathscr{D}_i^f \triangleq \zeta_i^T(t_f) Q_i^f \zeta_i(t_f) \times \underbrace{0 \times \cdots \times 0}_{(k_i-1)\text{-times}} .$$

Once immediate neighbors $j \in N_i$ of agent i fix the corresponding person-by-person equilibrium strategies u_j^* and thus the signaling or coordination effects u_{-i}^*, agent i then obtains an optimal stochastic control problem with risk-averse performance considerations. The construction of agent i's person-by-person policy now involves the 4-tuple $(K_{x_i},\ K_{z_i}, p_{x_i}, p_{z_i})$. Furthermore, the solutions of the equations (6.33)-(6.35) also depend on the admissible feedback gains K_{x_i} and K_{z_i}, in addition with the affine inputs p_{x_i} and p_{z_i}. In the sequel and elsewhere, when this dependence is needed to be clear, then the notations $\mathscr{H}_i(s, K_{x_i}, K_{z_i}; u_{-i}^*)$, $\breve{\mathscr{D}}_i(s, K_{x_i}, K_{z_i}, p_{x_i}, p_{z_i}; u_{-i}^*)$ and $\mathscr{D}_i(s, K_{x_i}, K_{z_i}, p_{x_i}, p_{z_i}; u_{-i}^*)$ should be used to denote the solution trajectories of the dynamics (6.33)-(6.35) with the admissible 5-tuple $(K_{x_i},\ K_{z_i}, p_{x_i}, p_{z_i}; u_{-i}^*)$.

For the given terminal data $(t_f, \mathscr{H}_i^f, \breve{\mathscr{D}}_i^f, \mathscr{D}_i^f)$, the theoretical framework for risk-averse control of the distributed stochastic system with possibly noncooperative u_{-i}^*, is then analyzed by a class of admissible feedback policies employed by agent i.

Definition 1 *Admissible Feedback Policies.*
Let compact subsets $\overline{K}^{x_i} \subset \mathbb{R}^{m_i \times n_i}$, $\overline{K}^{z_i} \subset \mathbb{R}^{q_i \times n_i}$, $\overline{P}^{x_i} \subset \mathbb{R}^{m_i}$ and $\overline{P}^{z_i} \subset \mathbb{R}^{q_i}$ be the sets of allowable feedback form values available at agent i and $i \in \overline{I}$. For the

given $k_i \in \mathbb{N}$ and sequence $\mu_i = \{\mu_r^i \geq 0\}_{r=1}^{k_i}$ with $\mu_1^i > 0$, the set of feedback gains $\mathscr{K}^{x_i}_{t_f,\mathscr{H}_i^f,\breve{\mathscr{D}}_i^f,\mathscr{D}_i^f;\mu_i}$, $\mathscr{K}^{z_i}_{t_f,\mathscr{H}_i^f,\breve{\mathscr{D}}_i^f,\mathscr{D}_i^f;\mu_i}$, $\mathscr{P}^{x_i}_{t_f,\mathscr{H}_i^f,\breve{\mathscr{D}}_i^f,\mathscr{D}_i^f;\mu_i}$ and $\mathscr{P}^{z_i}_{t_f,\mathscr{H}_i^f,\breve{\mathscr{D}}_i^f,\mathscr{D}_i^f;\mu_i}$ are assumed to be the classes of $\mathscr{C}(t_0,t_f;\mathbb{R}^{m_i\times n_i})$, $\mathscr{C}(t_0,t_f;\mathbb{R}^{q_i\times n_i})$, $\mathscr{C}(t_0,t_f;\mathbb{R}^{m_i})$ and $\mathscr{C}(t_0,t_f;\mathbb{R}^{q_i})$ with values $K_{x_i}(\cdot) \in \overline{K}^{x_i}$, $K_{z_i}(\cdot) \in \overline{K}^{z_i}$, $p_{x_i}(\cdot) \in \overline{P}^{x_i}$ and $p_{z_i}(\cdot) \in \overline{P}^{z_i}$, for which the solutions to the dynamic equations (6.33)-(6.35) with the terminal-value conditions $\mathscr{H}_i(t_f) = \mathscr{H}_i^f$, $\breve{\mathscr{D}}_i(t_f) = \breve{\mathscr{D}}_i^f$ and $\mathscr{D}_i(t_f) = \mathscr{D}_i^f$ exist on the interval of optimization $[t_0,t_f]$.

To determine agent i's the person-by-person equilibrium strategy with risk bearing so as to minimize its performance vulnerability of (6.12) against all the sample-path realizations from uncertain environments w_{ai} and noncooperative coordination u_{-i}^* from immediate neighbors j and $j \in N_i$, performance risks are henceforth interpreted as worries and fears about certain undesirable characteristics of performance distributions of (6.12) and thus are proposed to manage through a finite set of selective weights. This custom set of design freedoms representing particular uncertainty aversions is hence different from the ones with aversion to risk captured in risk-sensitive optimal control [9] and [10].

On $\mathscr{K}^{x_i}_{t_f,\mathscr{H}_i^f,\breve{\mathscr{D}}_i^f,\mathscr{D}_i^f;\mu_i}$, $\mathscr{K}^{z_i}_{t_f,\mathscr{H}_i^f,\breve{\mathscr{D}}_i^f,\mathscr{D}_i^f;\mu_i}$, $\mathscr{P}^{x_i}_{t_f,\mathscr{H}_i^f,\breve{\mathscr{D}}_i^f,\mathscr{D}_i^f;\mu_i}$ and $\mathscr{P}^{z_i}_{t_f,\mathscr{H}_i^f,\breve{\mathscr{D}}_i^f,\mathscr{D}_i^f;\mu_i}$, the performance index with risk-value considerations in risk-averse decision making is subsequently defined as follows.

Definition 2 *Risk-Value Aware Performance Index.*
Let incumbent agent i and $i \in \bar{I}$ select $k_i \in \mathbb{N}$ and the sequence of scalar coefficients $\mu_i = \{\mu_r^i \geq 0\}_{r=1}^{k_i}$ with $\mu_1^i > 0$. Then for the given initial condition (t_0, x_i^0), the risk-value aware performance index, $\phi_i^0 : \{t_0\} \times (\mathbb{R}^{n_i\times n_i})^{k_i} \times (\mathbb{R}^{n_i})^{k_i} \times \mathbb{R}^{k_i} \mapsto \mathbb{R}^+$ pertaining to risk-averse decision making of agent i over $[t_0,t_f]$ is defined by

$$\phi_i^0(t_0, \mathscr{H}_i(t_0), \breve{\mathscr{D}}_i(t_0), \mathscr{D}_i(t_0)) \triangleq \underbrace{\mu_1^i \kappa_1^i}_{\text{Value Measure}} + \underbrace{\mu_2^i \kappa_2^i + \cdots + \mu_{k_i}^i \kappa_{k_i}^i}_{\text{Risk Measures}}$$

$$= \sum_{r=1}^{k_i} \mu_r^i [(x_i^0)^T \mathscr{H}_i^r(t_0) x_i^0 + 2(x_i^0)^T \breve{\mathscr{D}}_i^r(t_0) + \mathscr{D}_i^r(t_0)], \quad (6.36)$$

wherein the additional design freedom by means of μ_r^i's utilized by agent i with risk-averse attitudes are sufficient to meet and exceed different levels of performance-based reliability requirements, for instance, mean (i.e., the average of performance measure), variance (i.e., the dispersion of values of performance measure around its mean), skewness (i.e., the anti-symmetry of the density of performance measure), kurtosis (i.e., the heaviness in the density tails of performance measure), etc., pertaining to closed-loop performance variations and uncertainties while the supporting solutions $\{\mathscr{H}_i^r(s)\}_{r=1}^{k_i}$, $\{\breve{\mathscr{D}}_i^r(s)\}_{r=1}^{k_i}$ and $\{\mathscr{D}_i^r(s)\}_{r=1}^{k_i}$ evaluated at $s = t_0$ satisfy the dynamical equations (6.33)-(6.35).

To specifically indicate the dependence of the risk-value aware performance index (6.36) expressed in Mayer form on $(K_{x_i}, K_{z_i}, p_{x_i}, p_{z_i})$ and the signaling effects u_{-i}^* issued by all immediate neighbors j from N_i, the multi-attribute utility function or performance index (6.36) for agent i is now rewritten explicitly as $\phi_i^0(K_{x_i}, K_{z_i}, p_{x_i}, p_{z_i}; u_{-i}^*)$.

Definition 3 *Nash Equilibrium Solution.*
An N-tuple of policies $\{(K_{x_1}^*, K_{z_1}^*, p_{x_1}^*, p_{z_1}^*), \ldots, (K_{x_N}^*, K_{z_N}^*, p_{x_N}^*, p_{z_N}^*)\}$ *is said to constitute a Nash equilibrium solution for the distributed N-agent stochastic game if, for all* $i \in \overline{N}$, *the Nash inequality condition holds*

$$\phi_i^0(K_{x_1}^*, K_{z_1}^*, p_{x_1}^*, p_{z_1}^*; u_{-i}^*) \le \phi_i^0(K_{x_1}, K_{z_1}, p_{x_1}, p_{z_1}; u_{-i}^*). \tag{6.37}$$

For the sake of time consistency and subgame perfection, a Nash equilibrium solution is required to have an additional property that its restriction on the interval $[t_0, \tau]$ is also a Nash solution to the truncated version of the original problem, defined on $[t_0, \tau]$. With such a restriction so defined, the Nash equilibrium solution is now termed as a feedback Nash equilibrium solution, which is now free of any informational nonuniqueness, and thus whose derivation allows a dynamic programming type argument.

Definition 4 *Feedback Nash Equilibrium.*
Let $(K_{x_i}^*, K_{z_i}^*, p_{x_i}^*, p_{z_i}^*)$ *constitute a feedback Nash strategy for agent i such that*

$$\phi_i^0(K_{x_i}^*, K_{z_i}^*, p_{x_i}^*, p_{z_i}^*; u_{-i}^*) \le \phi_i^0(K_{x_i}, K_{z_i}, p_{x_i}, p_{z_i}; u_{-i}^*), \qquad i \in \overline{I} \tag{6.38}$$

for admissible $K_{x_i} \in \mathscr{K}^{x_i}_{t_f, \mathscr{H}_i^f, \breve{\mathscr{D}}_i^f, \mathscr{D}_i^f; \mu_i}$, $K_{z_i} \in \mathscr{K}^{z_i}_{t_f, \mathscr{H}_i^f, \breve{\mathscr{D}}_i^f, \mathscr{D}_i^f; \mu_i}$, $p_{x_i} \in \mathscr{P}^{x_i}_{t_f, \mathscr{H}_i^f, \breve{\mathscr{D}}_i^f, \mathscr{D}_i^f; \mu_i}$ *and* $p_{z_i} \in \mathscr{P}^{z_i}_{t_f, \mathscr{H}_i^f, \breve{\mathscr{D}}_i^f, \mathscr{D}_i^f; \mu_i}$, *upon which the solutions to the dynamical systems (6.33)-(6.35) exist on* $[t_0, t_f]$.

Then, $\{(K_{x_1}^*, K_{z_1}^*, p_{x_1}^*, p_{z_1}^*), \ldots, (K_{x_N}^*, K_{z_N}^*, p_{x_N}^*, p_{z_N}^*)\}$ *when restricted to the interval* $[t_0, \tau]$ *is still a N-tuple feedback Nash equilibrium solution for the multiperson Nash decision problem with the appropriate terminal-value condition* $(\tau, \mathscr{H}_i^*(\tau), \breve{\mathscr{D}}_i^*(\tau), \mathscr{D}_i^*(\tau))$ *for all* $\tau \in [t_0, t_f]$.

In conformity with the rigorous formulation of dynamic programming, the following development is important. Let the terminal time t_f and 3-tuple states $(\mathscr{H}_i^f, \breve{\mathscr{D}}_i^f, \mathscr{D}_i^f)$, the other end condition involved the initial time t_0 and 3-tuple states $(\mathscr{H}_i^0, \breve{\mathscr{D}}_i^0, \mathscr{D}_i^0)$ be specified by a target set requirement.

Definition 5 *Target Sets.*
$(t_0, \mathscr{H}_i^0, \breve{\mathscr{D}}_i^0, \mathscr{D}_i^0) \in \mathscr{M}_i$, *where the* target set $\mathscr{M}_i$ *is a closed subset of* $[t_0, t_f] \times (\mathbb{R}^{n_i \times n_i})^{k_i} \times (\mathbb{R}^{n_i})^{k_i} \times \mathbb{R}^{k_i}$.

Now, the decision optimization residing at incumbent agent i is to minimize the risk-value aware performance index (6.36) over admissible feedback strategies composed by $K_{x_i} \equiv K_{x_i}(\cdot) \in \mathscr{K}^{x_i}_{t_f, \mathscr{H}_i^f, \breve{\mathscr{D}}_i^f, \mathscr{D}_i^f; \mu_i}$, $K_{z_i} \equiv K_{z_i}(\cdot) \in \mathscr{K}^{z_i}_{t_f, \mathscr{H}_i^f, \breve{\mathscr{D}}_i^f, \mathscr{D}_i^f; \mu_i}$, $p_{x_i} \equiv p_{x_i}(\cdot) \in \mathscr{P}^{x_i}_{t_f, \mathscr{H}_i^f, \breve{\mathscr{D}}_i^f, \mathscr{D}_i^f; \mu_i}$ and $p_{z_i} \equiv p_{z_i}(\cdot) \in \mathscr{P}^{z_i}_{t_f, \mathscr{H}_i^f, \breve{\mathscr{D}}_i^f, \mathscr{D}_i^f; \mu_i}$ while subject to interconnection links from all immediate neighbors with corresponding feedback Nash policies u_{-i}^*.

Definition 6 *Optimization of Mayer Problem.*
Given the sequence of scalars $\mu_i = \{\mu_r^i \geq 0\}_{r=1}^{k^i}$ with $\mu_1^i > 0$, the decision optimization over $[t_0, t_f]$ at agent i and $i \in \bar{I}$ is given by

$$\min_{K_{x_i}, K_{z_i}, p_{x_i}, p_{z_i}} \phi_i^0(K_{x_i}, K_{z_i}, p_{x_i}, p_{z_i}; u_{-i}^*) \tag{6.39}$$

subject to the dynamical equations (6.33)-(6.35) on $[t_0, t_f]$.

Notice that the optimization considered here is in Mayer form and can be solved by applying an adaptation of the Mayer form verification results as given in [11]. To embed this optimization facing agent i into a larger problem, the terminal time and states $(t_f, \mathscr{H}_i^f, \breve{\mathscr{D}}_i^f, \mathscr{D}_i^f)$ are parameterized as $(\varepsilon, \mathscr{Y}_i, \breve{\mathscr{Z}}_i, \mathscr{Z}_i)$, whereby $\mathscr{Y}_i \triangleq \mathscr{H}_i(\varepsilon)$, $\breve{\mathscr{Z}}_i \triangleq \breve{\mathscr{D}}_i(\varepsilon)$ and $\mathscr{Z}_i \triangleq \mathscr{D}_i(\varepsilon)$. Thus, the value function for this optimization problem is now depending on the parameterization of terminal-value conditions.

Definition 7 *Value Function.*
Suppose $(\varepsilon, \mathscr{Y}_i, \breve{\mathscr{Z}}_i, \mathscr{Z}_i) \in [t_0, t_f] \times (\mathbb{R}^{n_i \times n_i})^{k_i} \times (\mathbb{R}^{n_i})^{k_i} \times \mathbb{R}^{k_i}$ is given and fixed for agent i and $i \in \bar{I}$. Then, the value function $\mathscr{V}_i(\varepsilon, \mathscr{Y}_i, \breve{\mathscr{Z}}_i, \mathscr{Z}_i)$ is defined by

$$\mathscr{V}_i(\varepsilon, \mathscr{Y}_i, \breve{\mathscr{Z}}_i, \mathscr{Z}_i) \triangleq \inf_{K_{x_i}, K_{z_i}, p_{x_i}, p_{z_i}} \phi_i^0(K_{x_i}, K_{z_i}, p_{x_i}, p_{z_i}; u_{-i}^*) .$$

For convention, $\mathscr{V}_i(\varepsilon, \mathscr{Y}_i, \breve{\mathscr{Z}}_i, \mathscr{Z}_i) \triangleq \infty$ when $\mathscr{K}^{x_i}_{t_f, \mathscr{H}_i^f, \breve{\mathscr{D}}_i^f, \mathscr{D}_i^f; \mu_i} \times \mathscr{K}^{z_i}_{t_f, \mathscr{H}_i^f, \breve{\mathscr{D}}_i^f, \mathscr{D}_i^f; \mu_i} \times \mathscr{P}^{x_i}_{t_f, \mathscr{H}_i^f, \breve{\mathscr{D}}_i^f, \mathscr{D}_i^f; \mu_i} \times \mathscr{P}^{z_i}_{t_f, \mathscr{H}_i^f, \breve{\mathscr{D}}_i^f, \mathscr{D}_i^f; \mu_i}$ is empty. Next, some candidates for the value function are constructed with the help of the concept of reachable set.

Definition 8 *Reachable Sets.*
Associate with agent i and $i \in \bar{I}$ a reachable set defined by $\mathscr{Q}_i \triangleq \Big\{(\varepsilon, \mathscr{Y}_i, \breve{\mathscr{Z}}_i, \mathscr{Z}_i) \in [t_0, t_f] \times (\mathbb{R}^{n_i \times n_i})^{k_i} \times (\mathbb{R}^{n_i})^{k_i} \times \mathbb{R}^{k_i}$ such that the Cartesian product $\mathscr{K}^{x_i}_{t_f, \mathscr{H}_i^f, \breve{\mathscr{D}}_i^f, \mathscr{D}_i^f; \mu_i} \times \mathscr{K}^{z_i}_{t_f, \mathscr{H}_i^f, \breve{\mathscr{D}}_i^f, \mathscr{D}_i^f; \mu_i} \times \mathscr{P}^{x_i}_{t_f, \mathscr{H}_i^f, \breve{\mathscr{D}}_i^f, \mathscr{D}_i^f; \mu_i} \times \mathscr{P}^{z_i}_{t_f, \mathscr{H}_i^f, \breve{\mathscr{D}}_i^f, \mathscr{D}_i^f; \mu_i} \neq \emptyset \Big\}$.

Moreover, it can be shown that the value function associated with agent i is satisfying a partial differential equation at interior points of $\mathscr{Q}_i$, at which it is differentiable.

Theorem 3 *Hamilton-Jacobi-Bellman (HJB) Equation-Mayer Problem.*
Let $(\varepsilon, \mathscr{Y}_i, \breve{\mathscr{Z}}_i, \mathscr{Z}_i)$ be any interior point of the reachable set $\mathscr{Q}_i$, at which the value function $\mathscr{V}_i(\varepsilon, \mathscr{Y}_i, \breve{\mathscr{Z}}_i, \mathscr{Z}_i)$ is differentiable. If there exists a feedback Nash strategy which is supported by $K_{x_i}^*(\cdot) \in \mathscr{K}^{x_i}_{t_f, \mathscr{H}_i^f, \breve{\mathscr{D}}_i^f, \mathscr{D}_i^f; \mu_i}$, $K_{z_i}^*(\cdot) \in \mathscr{K}^{z_i}_{t_f, \mathscr{H}_i^f, \breve{\mathscr{D}}_i^f, \mathscr{D}_i^f; \mu_i}$, $p_{x_i}^*(\cdot) \in \mathscr{P}^{x_i}_{t_f, \mathscr{H}_i^f, \breve{\mathscr{D}}_i^f, \mathscr{D}_i^f; \mu_i}$ and $p_{z_i}^*(\cdot) \in \mathscr{P}^{z_i}_{t_f, \mathscr{H}_i^f, \breve{\mathscr{D}}_i^f, \mathscr{D}_i^f; \mu_i}$, then the differential equation

$$
\begin{aligned}
0 = \min_{K_{x_i}\in\overline{K}^{x_i}, K_{z_i}\in\overline{K}^{z_i}, p_{x_i}\in\overline{P}^{x_i}, p_{z_i}\in\overline{P}^{z_i}} & \left\{ \frac{\partial}{\partial\varepsilon}\mathscr{V}_i(\varepsilon,\mathscr{Y}_i,\breve{\mathscr{Z}}_i,\mathscr{Z}_i) \right. \\
& + \frac{\partial}{\partial\,\mathrm{vec}(\mathscr{Y}_i)}\mathscr{V}_i(\varepsilon,\mathscr{Y}_i,\breve{\mathscr{Z}}_i,\mathscr{Z}_i)\mathrm{vec}(\mathscr{F}_i(\varepsilon,\mathscr{Y}_i,K_{x_i},K_{z_i})) \\
& + \frac{\partial}{\partial\,\mathrm{vec}(\breve{\mathscr{Z}}_i)}\mathscr{V}_i(\varepsilon,\mathscr{Y}_i,\breve{\mathscr{Z}}_i,\mathscr{Z}_i)\mathrm{vec}(\breve{\mathscr{G}}_i(\varepsilon,\mathscr{Y}_i,\breve{\mathscr{Z}}_i,K_{x_i},K_{z_i},p_{x_i},p_{z_i}) \\
& \left. + \frac{\partial}{\partial\,\mathrm{vec}(\mathscr{Z}_i)}\mathscr{V}_i(\varepsilon,\mathscr{Y}_i,\breve{\mathscr{Z}}_i,\mathscr{Z}_i)\mathrm{vec}(\mathscr{G}_i(\varepsilon,\mathscr{Y}_i,\breve{\mathscr{Z}}_i,p_{x_i},p_{z_i}) \right\}
\end{aligned}
\tag{6.40}
$$

is satisfied whereby $\mathscr{V}_i(t_0,\mathscr{Y}_i(t_0),\breve{\mathscr{Z}}_i(t_0),\mathscr{Z}_i(t_0)) = \phi_i^0(\mathscr{H}_i(t_0),\breve{\mathscr{D}}_i(t_0),\mathscr{D}_i(t_0))$.

Proof. By what have been shown in the recent results by the first author [7], the proof for the result herein is readily proven.

Finally, the following result gives the sufficient condition used to verify a feedback Nash strategy for incumbent agent i and $i \in \overline{I}$.

Theorem 4 *Verification Theorem.*
Let $\mathscr{W}_i(\varepsilon,\mathscr{Y}_i,\breve{\mathscr{Z}}_i,\mathscr{Z}_i)$ *be continuously differentiable solution of the HJB equation (6.40) for agent i and* $i\in\overline{I}$, *which satisfies the boundary condition*

$$
\mathscr{W}_i(t_0,\mathscr{H}_i(t_0),\breve{\mathscr{D}}_i(t_0),\mathscr{D}_i(t_0)) = \phi_i^0(t_0,\mathscr{H}_i(t_0),\breve{\mathscr{D}}_i(t_0),\mathscr{D}_i(t_0)) \ . \tag{6.41}
$$

Let $(t_f,\mathscr{H}_i^f,\breve{\mathscr{D}}_i^f,\mathscr{D}_i^f)$ *be a 4-tuple point in* $\mathscr{Q}_i$; *let* $K_{x_i}\in\mathscr{K}^{x_i}_{t_f,\mathscr{H}_i^f,\breve{\mathscr{D}}_i^f,\mathscr{D}_i^f;\mu_i}$, $K_{z_i}\in\mathscr{K}^{z_i}_{t_f,\mathscr{H}_i^f,\breve{\mathscr{D}}_i^f,\mathscr{D}_i^f;\mu_i}$, $p_{x_i}\in\mathscr{P}^{x_i}_{t_f,\mathscr{H}_i^f,\breve{\mathscr{D}}_i^f,\mathscr{D}_i^f;\mu_i}$, $p_{z_i}\in\mathscr{P}^{z_i}_{t_f,\mathscr{H}_i^f,\breve{\mathscr{D}}_i^f,\mathscr{D}_i^f;\mu_i}$; *and let* $\mathscr{H}_i(\cdot)$, $\breve{\mathscr{D}}_i(\cdot)$ *and* $\mathscr{D}_i(\cdot)$ *be the corresponding solutions of the equations of motion (6.33)-(6.35). Then,* $\mathscr{W}_i(s,\mathscr{H}_i(s),\breve{\mathscr{D}}_i(s),\mathscr{D}_i(s))$ *is time-backward increasing function of s.*

If $K^*_{x_i}\in\mathscr{K}^{x_i}_{t_f,\mathscr{H}_i^f,\breve{\mathscr{D}}_i^f,\mathscr{D}_i^f;\mu_i}$, $K^*_{z_i}\in\mathscr{K}^{z_i}_{t_f,\mathscr{H}_i^f,\breve{\mathscr{D}}_i^f,\mathscr{D}_i^f;\mu_i}$, $p^*_{x_i}\in\mathscr{P}^{x_i}_{t_f,\mathscr{H}_i^f,\breve{\mathscr{D}}_i^f,\mathscr{D}_i^f;\mu_i}$ *and* $p^*_{z_i}\in\mathscr{P}^{z_i}_{t_f,\mathscr{H}_i^f,\breve{\mathscr{D}}_i^f,\mathscr{D}_i^f;\mu_i}$ *defining a person-by-person equilibrium or feedback Nash strategy for agent i and* $i\in\overline{I}$ *with the corresponding solutions* $\mathscr{H}_i^*(\cdot)$, $\breve{\mathscr{D}}_i^*(\cdot)$ *and* $\mathscr{D}_i^*(\cdot)$ *of the dynamical equations (6.33)-(6.35) such that, for* $s\in[t_0,t_f]$

$$
\begin{aligned}
0 = {} & \frac{\partial}{\partial\varepsilon}\mathscr{W}_i(s,\mathscr{H}_i^*(s),\breve{\mathscr{D}}_i^*(s),\mathscr{D}_i^*(s)) + \frac{\partial}{\partial\,\mathrm{vec}(\mathscr{Y}_i)}\mathscr{W}_i(s,\mathscr{H}_i^*(s),\breve{\mathscr{D}}_i^*(s),\mathscr{D}_i^*(s)) \\
& \cdot\mathrm{vec}(\mathscr{F}_i(s,\mathscr{Y}_i^*(s),K^*_{x_i}(s),K^*_{z_i}(s))) + \frac{\partial}{\partial\,\mathrm{vec}(\breve{\mathscr{Z}}_i)}\mathscr{W}_i(s,\mathscr{H}_i^*(s),\breve{\mathscr{D}}_i^*(s),\mathscr{D}_i^*(s)) \\
& \cdot\mathrm{vec}(\breve{\mathscr{G}}_i(s,\mathscr{H}_i^*(s),\breve{\mathscr{D}}_i^*(s),K^*_{x_i}(s),K^*_{z_i}(s),p^*_{x_i}(s),p^*_{z_i}(s)) \\
& + \frac{\partial}{\partial\,\mathrm{vec}(\mathscr{Z}_i)}\mathscr{W}_i(s,\mathscr{H}_i^*(s),\breve{\mathscr{D}}_i^*(s),\mathscr{D}_i^*(s))\mathrm{vec}(\mathscr{G}_i(s,\mathscr{H}_i^*(s),\breve{\mathscr{D}}_i^*(s),p^*_{x_i}(s),p^*_{z_i}(s))
\end{aligned}
\tag{6.42}
$$

then $(K_{x_i}^*, K_{z_i}^*, p_{x_i}^*, p_{z_i}^*)$ *results in a feedback Nash strategy or person-by-person equilibrium for agent i in* $\mathscr{K}^{x_i}_{t_f,\mathscr{H}_i^f,\breve{\mathscr{D}}_i^f,\mathscr{D}_i^f;\mu_i} \times \mathscr{K}^{z_i}_{t_f,\mathscr{H}_i^f,\breve{\mathscr{D}}_i^f,\mathscr{D}_i^f;\mu_i} \times \mathscr{P}^{x_i}_{t_f,\mathscr{H}_i^f,\breve{\mathscr{D}}_i^f,\mathscr{D}_i^f;\mu_i} \times \mathscr{P}^{z_i}_{t_f,\mathscr{H}_i^f,\breve{\mathscr{D}}_i^f,\mathscr{D}_i^f;\mu_i}$. *Furthermore, it follows that*

$$\mathscr{W}_i(\varepsilon, \mathscr{Y}_i, \breve{\mathscr{Z}}_i, \mathscr{Z}_i) = \mathscr{V}_i(\varepsilon, \mathscr{Y}_i, \breve{\mathscr{Z}}_i, \mathscr{Z}_i), \tag{6.43}$$

whereby $\mathscr{V}_i(\varepsilon, \mathscr{Y}_i, \breve{\mathscr{Z}}_i, \mathscr{Z}_i)$ *is the value function associated with incumbent agent i.*

Proof. With the aid of the recent development [7], the proof then follows for the verification theorem herein.

6.4 Distributed Person-by-Person Equilibrium Strategies

Reflecting on the Mayer-form optimization problem of the person-by-person equilibrium strategy concerned by incumbent agent i and $i \in \bar{I}$, the technical approach is to apply an adaptation of the Mayer-form verification theorem of dynamic programming as given in [11]. In the framework of dynamic programming, it is often required to denote the terminal time and states of a family of optimization problems as $(\varepsilon, \mathscr{Y}_i, \breve{\mathscr{Z}}_i, \mathscr{Z}_i)$ rather than $(t_f, \mathscr{H}_i^f, \breve{\mathscr{D}}_i^f, \mathscr{D}_i^f)$. Stating precisely, for $\varepsilon \in [t_0, t_f]$ and $1 \le r \le k_i$, the states of the performance robustness (6.33)-(6.35) defined on the interval $[t_0, \varepsilon]$ have the terminal values denoted by $\mathscr{H}_i(\varepsilon) \equiv \mathscr{Y}_i$, $\breve{\mathscr{D}}_i(\varepsilon) \equiv \breve{\mathscr{Z}}_i$ and $\mathscr{D}_i(\varepsilon) \equiv \mathscr{Z}_i$.

Since the performance index (6.36) is quadratic affine in terms of arbitrarily fixed x_i^0, the resulting insight suggests a solution to the adapted HJB equation (6.40) is of the form as follows. It is assumed that $(\varepsilon, \mathscr{Y}_i, \breve{\mathscr{Z}}_i, \mathscr{Z}_i)$ is any interior point of the reachable set $\mathscr{Q}_i$ at which the real-valued function

$$\begin{aligned}\mathscr{W}_i(\varepsilon, \mathscr{Y}_i, \breve{\mathscr{Z}}_i, \mathscr{Z}_i) &= (x_i^0)^T \sum_{r=1}^{k_i} \mu_r^i (\mathscr{Y}_i^r + \mathscr{E}_i^r(\varepsilon)) x_i^0 \\ &\quad + 2(x_i^0)^T \sum_{r=1}^{k_i} \mu_r^i (\breve{\mathscr{Z}}_i^r + \breve{\mathscr{T}}_i^r(\varepsilon)) + \sum_{r=1}^{k_i} \mu_r^i (\mathscr{Z}_i^r + \mathscr{T}_i^r(\varepsilon)) \end{aligned} \tag{6.44}$$

is differentiable. The parametric functions of time $\mathscr{E}_i^r \in \mathscr{C}^1(t_0, t_f; \mathbb{R}^{n_i \times n_i})$, $\breve{\mathscr{T}}_i^r \in \mathscr{C}^1(t_0, t_f; \mathbb{R}_i^n)$ and $\mathscr{T}_i^r \in \mathscr{C}^1(t_0, t_f; \mathbb{R})$ are yet to be determined. Furthermore, the time derivative of $\mathscr{W}_i(\varepsilon, \mathscr{Y}_i, \breve{\mathscr{Z}}_i, \mathscr{Z}_i)$ can be shown to be

$$\begin{aligned}\frac{d}{d\varepsilon}\mathscr{W}_i(\varepsilon, \mathscr{Y}_i, \breve{\mathscr{Z}}_i, \mathscr{Z}_i) &= (x_i^0)^T \sum_{r=1}^{k_i} \mu_r^i \Big(\mathscr{F}_i^r(\varepsilon, \mathscr{Y}_i, K_{x_i}, K_{z_i}) + \frac{d}{d\varepsilon}\mathscr{E}_i^r(\varepsilon)\Big) x_i^0 \\ &\quad + 2(x_i^0)^T \sum_{r=1}^{k_i} \mu_r^i \Big(\breve{\mathscr{G}}_i^r(\varepsilon, \mathscr{Y}_i, \breve{\mathscr{Z}}_i, K_{x_i}, K_{z_i}, p_{x_i}, p_{z_i}) + \frac{d}{d\varepsilon}\breve{\mathscr{T}}_i^r(\varepsilon)\Big) \\ &\quad + \sum_{r=1}^{k_i} \mu_r^i \Big(\mathscr{G}_i^r(\varepsilon, \mathscr{Y}_i, \breve{\mathscr{Z}}_i, p_{x_i}, p_{z_i}) + \frac{d}{d\varepsilon}\mathscr{T}_i^r(\varepsilon)\Big). \end{aligned} \tag{6.45}$$

The substitution of this hypothesized solution (6.44) into the HJB equation (6.40) and making use of (6.45) results in

$$0 \equiv \min_{K_{x_i}\in\overline{K}^{x_i}, K_{z_i}\in\overline{K}^{z_i}, p_{x_i}\in\overline{P}^{x_i}, p_{z_i}\in\overline{P}^{z_i}} \Big\{ (x_i^0)^T \sum_{r=1}^{k_i} \mu_r^i \frac{d}{d\varepsilon} \mathscr{E}_i^r(\varepsilon) x_i^0 + 2(x_i^0)^T \sum_{r=1}^{k_i} \mu_r^i \frac{d}{d\varepsilon} \breve{\mathscr{T}}_i^r(\varepsilon)$$
$$+ \sum_{r=1}^{k_i} \mu_r^i \frac{d}{d\varepsilon} \mathscr{T}_i^r(\varepsilon) + 2(x_i^0)^T \sum_{r=1}^{k_i} \mu_r^i \breve{\mathscr{G}}_i^r(\varepsilon, \mathscr{Y}_i, \breve{\mathscr{Z}}_i, K_{x_i}, K_{z_i}, p_{x_i}, p_{z_i})$$
$$+ (x_i^0)^T \sum_{r=1}^{k_i} \mu_r^i \mathscr{F}_i^r(\varepsilon, \mathscr{Y}_i, K_{x_i}, K_{z_i}) x_i^0 + \sum_{r=1}^{k_i} \mu_r^i \mathscr{G}_i^r(\varepsilon, \mathscr{Y}_i, \breve{\mathscr{Z}}_i, p_{x_i}, p_{z_i}) \Big\}. \quad (6.46)$$

Differentiating the expression within the bracket of (6.46) with respect to K_{x_i}, K_{z_i}, p_{x_i} and p_{z_i} yields the necessary conditions for an extremum of (6.40) on $[t_0, \varepsilon]$,

$$0 \equiv -2[B_{ii}^T(\varepsilon) \sum_{r=1}^{k_i} \mu_r^i \mathscr{Y}_i^r + \mu_i^1 R_{ii}(\varepsilon) K_{x_i}] x_i^0 (x_i^0)^T$$
$$- 2[B_{ii}^T(\varepsilon) \sum_{r=1}^{k_i} \mu_i^r \breve{\mathscr{Z}}_i^r + \mu_i^1 R_{ii}(\varepsilon) p_{x_i}] (x_i^0)^T$$
$$0 \equiv -2[C_{ii}^T(\varepsilon) \sum_{r=1}^{k_i} \mu_i^r \mathscr{Y}_i^r + \mu_i^1 R_{zi}(\varepsilon) K_{z_i}] x_i^0 (x_i^0)^T$$
$$- 2[C_{ii}^T(\varepsilon) \sum_{r=1}^{k_i} \mu_i^r \breve{\mathscr{Z}}_i^r + \mu_i^1 R_{zi}(\varepsilon) p_{z_i}] (x_i^0)^T$$
$$0 \equiv -2[B_{ii}^T(\varepsilon) \sum_{r=1}^{k_i} \mu_i^r \mathscr{Y}_i^r + \mu_i^1 R_{ii}(\varepsilon) K_{x_i}] x_i^0$$
$$- 2[B_{ii}^T(\varepsilon) \sum_{r=1}^{k_i} \mu_i^r \breve{\mathscr{Z}}_i^r + \mu_i^1 R_{ii}(\varepsilon) p_{x_i}]$$
$$0 \equiv -2[C_{ii}^T(\varepsilon) \sum_{r=1}^{k_i} \mu_i^r \mathscr{Y}_i^r + \mu_i^1 R_{zi}(\varepsilon) K_{z_i}] x_i^0$$
$$- 2[C_{ii}^T(\varepsilon) \sum_{r=1}^{k_i} \mu_i^r \breve{\mathscr{Z}}_i^r + \mu_i^1 R_{zi}(\varepsilon) p_{z_i}]$$

Because all $x_i^0 (x_i^0)^T$, $(x_i^0)^T$ and x_i^0 have arbitrary ranks of one, it must be true that

$$K_{x_i} = -(\mu_i^1 R_{ii}(\varepsilon))^{-1} B_{ii}^T(\varepsilon) \sum_{r=1}^{k_i} \mu_i^r \mathscr{Y}_i^r, \quad (6.47)$$

$$K_{z_i} = -(\mu_i^1 R_{zi}(\varepsilon))^{-1} C_{ii}^T(\varepsilon) \sum_{r=1}^{k_i} \mu_i^r \mathscr{Y}_i^r , \tag{6.48}$$

$$p_{x_i} = -(\mu_i^1 R_{ii}(\varepsilon))^{-1} B_{ii}^T(\varepsilon) \sum_{r=1}^{k_i} \mu_i^r \breve{\mathscr{Z}}_i^r , \tag{6.49}$$

$$p_{z_i} = -(\mu_i^1 R_{zi}(\varepsilon))^{-1} C_{ii}^T(\varepsilon) \sum_{r=1}^{k_i} \mu_i^r \breve{\mathscr{Z}}_i^r . \tag{6.50}$$

Replacing these results (6.47)-(6.50) into the right member of the HJB equation (6.40) yields the value of the minimum

$$
\begin{aligned}
&(x_i^0)^T \sum_{r=1}^{k_i} \mu_i^r \frac{d}{d\varepsilon} \mathscr{E}_i^r(\varepsilon) x_i^0 + 2(x_i^0)^T \sum_{r=1}^{k_i} \mu_i^r \frac{d}{d\varepsilon} \breve{\mathscr{T}}_i^r(\varepsilon) + \sum_{r=1}^{k_i} \mu_i^r \frac{d}{d\varepsilon} \mathscr{T}_i^r(\varepsilon) \\
&\quad + (x_i^0)^T \Big\{ - \big[A_{ii}(\varepsilon) - B_{ii}(\varepsilon)(\mu_i^1 R_{ii}(\varepsilon))^{-1} B_{ii}^T(\varepsilon) \sum_{s=1}^{k_i} \mu_i^s \mathscr{Y}_i^s \\
&\quad - C_{ii}(\varepsilon)(\mu_i^1 R_{zi}(\varepsilon))^{-1} C_{ii}^T(\varepsilon) \sum_{s=1}^{k_i} \mu_i^s \mathscr{Y}_i^s \big]^T \sum_{r=1}^{k_i} \mu_i^r \mathscr{Y}_i^r \\
&\quad - \sum_{r=1}^{k_i} \mu_i^r \mathscr{Y}_i^r \big[A_{ii}(\varepsilon) - B_{ii}(\varepsilon)(\mu_i^1 R_{ii}(\varepsilon))^{-1} B_{ii}^T(\varepsilon) \sum_{s=1}^{k_i} \mu_i^s \mathscr{Y}_i^s \\
&\quad - C_{ii}(\varepsilon)(\mu_i^1 R_{zi}(\varepsilon))^{-1} C_{ii}^T(\varepsilon) \sum_{s=1}^{k_i} \mu_i^s \mathscr{Y}_i^s \big] - \mu_i^1 Q_{ii}(\varepsilon) - \mu_i^1 Q_i(\varepsilon) \\
&\quad - \mu_i^1 \sum_{s=1}^{k_i} \mu_i^s \mathscr{Y}_i^s B_{ii}(\varepsilon)(\mu_i^1 R_{ii}(\varepsilon))^{-1} R_{ii}(\varepsilon)(\mu_i^1 R_{ii}(\varepsilon))^{-1} B_{ii}^T(\varepsilon) \sum_{r=1}^{k_i} \mu_i^r \mathscr{Y}_i^r \\
&\quad - \mu_i^1 \sum_{s=1}^{k_i} \mu_i^s \mathscr{Y}_i^s C_{ii}(\varepsilon)(\mu_i^1 R_{zi}(\varepsilon))^{-1} R_{zi}(\varepsilon)(\mu_i^1 R_{zi}(\varepsilon))^{-1} C_{ii}^T(\varepsilon) \sum_{r=1}^{k_i} \mu_i^r \mathscr{Y}_i^r \\
&\quad - \sum_{r=2}^{k_i} \mu_r^i \sum_{v=1}^{r-1} \frac{2r!}{v!(r-v)!} \mathscr{Y}_i^v G_i(\varepsilon) W_i G_i^T(\varepsilon) \mathscr{Y}_i^{r-v} \\
&\quad - \sum_{r=2}^{k_i} \mu_r^i \sum_{v=1}^{r-1} \frac{2r!}{v!(r-v)!} \mathscr{Y}_i^{k_i+v} G_{ic}(\varepsilon) V_c G_{ic}^T(\varepsilon) \mathscr{Y}_i^{2k_i+r-v} \Big\} x_i^0 \\
&\quad + 2(x_i^0)^T \Big\{ - \big[A_{ii}(\varepsilon) - B_{ii}(\varepsilon)(\mu_i^1 R_{ii}(\varepsilon))^{-1} B_{ii}^T(\varepsilon) \sum_{s=1}^{k_i} \mu_i^s \mathscr{Y}_i^s
\end{aligned}
$$

$$
\begin{aligned}
&-C_{ii}(\varepsilon)(\mu_i^1 R_{zi}(\varepsilon))^{-1}C_{ii}^T(\varepsilon)\sum_{s=1}^{k_i}\mu_i^s\mathscr{Y}_i^s\big]^T\sum_{r=1}^{k_i}\mu_i^r\mathscr{Z}_i^r\\
&\qquad-\sum_{r=1}^{k_i}\mu_i^r\mathscr{Y}_i^r\big[-B_{ii}(\varepsilon)(\mu_i^1 R_{ii}(\varepsilon))^{-1}B_{ii}^T(\varepsilon)\sum_{s=1}^{k_i}\mu_i^s\breve{\mathscr{Z}}_i^s\\
&\qquad-C_{ii}(\varepsilon)(\mu_i^1 R_{zi}(\varepsilon))^{-1}C_{ii}^T(\varepsilon)\sum_{s=1}^{k_i}\mu_i^s\breve{\mathscr{Z}}_i^s+\sum_{j=1}^{N_i}B_{ij}(\varepsilon)u_{ij}^*(\varepsilon)\big]\\
&\qquad-\sum_{r=1}^{k_i}\mu_i^r\mathscr{Y}_i^{k_i+r}B_{ic}(\varepsilon)u_{ic}(\varepsilon)+\mu_i^1Q_i(\varepsilon)\zeta_i(\varepsilon)\\
&\quad-\mu_i^1\sum_{s=1}^{k_i}\mu_i^s\mathscr{Y}_i^sB_{ii}(\varepsilon)(\mu_i^1R_{ii}(\varepsilon))^{-1}R_{ii}(\varepsilon)(\mu_i^1R_{ii}(\varepsilon))^{-1}B_{ii}^T(\varepsilon)\sum_{r=1}^{k_i}\mu_i^r\breve{\mathscr{Z}}_i^r\\
&\quad-\mu_i^1\sum_{s=1}^{k_i}\mu_i^s\mathscr{Y}_i^sC_{ii}(\varepsilon)(\mu_i^1R_{zi}(\varepsilon))^{-1}R_{zi}(\varepsilon)(\mu_i^1R_{zi}(\varepsilon))^{-1}C_{ii}^T(\varepsilon)\sum_{r=1}^{k_i}\mu_i^r\breve{\mathscr{Z}}_i^r\Big\}\\
&\qquad-\mathrm{Tr}\Big\{\sum_{r=1}^{k_i}\mu_i^r\mathscr{Y}_i^rG_i(\varepsilon)W_iG_i^T(\varepsilon)+\sum_{r=1}^{k_i}\mu_i^r\mathscr{Y}_i^{3k_i+r}G_{ic}(\varepsilon)V_cG_{ic}^T(\varepsilon)\Big\}\\
&\qquad-2\sum_{r=1}^{k_i}\mu_i^r(\breve{\mathscr{Z}}_i^r)^T\big[-B_{ii}(\varepsilon)(\mu_i^1R_{ii}(\varepsilon))^{-1}B_{ii}^T(\varepsilon)\sum_{s=1}^{k_i}\mu_i^s\breve{\mathscr{Z}}_i^s\\
&\qquad-C_{ii}(\varepsilon)(\mu_i^1R_{zi}(\varepsilon))^{-1}C_{ii}^T(\varepsilon)\sum_{s=1}^{k_i}\mu_i^s\breve{\mathscr{Z}}_i^s+\sum_{j=1}^{N_i}B_{ij}(\varepsilon)u_{ij}^*(\varepsilon)\big]\\
&\qquad-2\sum_{r=1}^{k_i}\mu_i^r(\breve{\mathscr{Z}}_i^{k_i+r})^TB_{ic}(\varepsilon)u_{ic}(\varepsilon)-\mu_i^1\zeta_i^T(\varepsilon)Q_i(\varepsilon)\zeta_i(\varepsilon)\\
&\quad-\mu_i^1\sum_{r=1}^{k_i}\mu_i^r(\breve{\mathscr{Z}}_i^r)^TB_{ii}(\varepsilon)(\mu_i^1R_{ii}(\varepsilon))^{-1}R_{ii}(\varepsilon)(\mu_i^1R_{ii}(\varepsilon))^{-1}B_{ii}^T(\varepsilon)\sum_{s=1}^{k_i}\mu_i^s\breve{\mathscr{Z}}_i^s\\
&-\mu_i^1\sum_{r=1}^{k_i}\mu_i^r(\breve{\mathscr{Z}}_i^r)^TC_{ii}(\varepsilon)(\mu_i^1R_{zi}(\varepsilon))^{-1}R_{zi}(\varepsilon)(\mu_i^1R_{zi}(\varepsilon))^{-1}C_{ii}^T(\varepsilon)\sum_{s=1}^{k_i}\mu_i^s\breve{\mathscr{Z}}_i^s\,. \qquad (6.51)
\end{aligned}
$$

For each agent i, it is necessary to exhibit $\{\mathscr{E}_i^r(\cdot)\}_{r=1}^{k_i}$, $\{\breve{\mathscr{T}}_i^r(\cdot)\}_{r=1}^{k_i}$ and $\{\mathscr{T}_i^r(\cdot)\}_{r=1}^{k_i}$ which render the left side of the HJB equation (6.40) equal to zero for $\varepsilon \in [t_0, t_f]$, when $\{\mathscr{Y}_i^r\}_{r=1}^{k_i}$, $\{\breve{\mathscr{Z}}_i^r\}_{r=1}^{k_i}$ and $\{\mathscr{Z}_i^r\}_{r=1}^{k_i}$ are evaluated along the solution trajectories of the dynamical equations (6.33)-(6.35). With a careful examination of the expression (6.51), it reveals that

$$
\begin{aligned}
\frac{d}{d\varepsilon}\mathscr{E}_i^1(\varepsilon) &= [A_{ii}(\varepsilon) - B_{ii}(\varepsilon)R_{ii}^{-1}(\varepsilon)B_{ii}^T(\varepsilon)\sum_{s=1}^{k_i}\frac{\mu_i^s}{\mu_i^1}\mathscr{H}_i^s(\varepsilon) \\
&\quad - C_{ii}(\varepsilon)R_{zi}^{-1}(\varepsilon)C_{ii}^T(\varepsilon)\sum_{s=1}^{k_i}\frac{\mu_i^s}{\mu_i^1}\mathscr{H}_i^s(\varepsilon)]^T\mathscr{H}_i^1(\varepsilon) \\
&\quad + \mathscr{H}_i^1(\varepsilon)\big[A_{ii}(\varepsilon) - B_{ii}(\varepsilon)R_{ii}^{-1}(\varepsilon)B_{ii}^T(\varepsilon)\sum_{s=1}^{k_i}\frac{\mu_i^s}{\mu_i^1}\mathscr{H}_i^s(\varepsilon) \\
&\quad - C_{ii}(\varepsilon)R_{zi}^{-1}(\varepsilon)C_{ii}^T(\varepsilon)\sum_{s=1}^{k_i}\frac{\mu_i^s}{\mu_i^1}\mathscr{H}_i^s(\varepsilon)\big] + Q_{ii}(\varepsilon) + Q_i(\varepsilon) \\
&\quad + \sum_{s=1}^{k_i}\frac{\mu_i^s}{\mu_i^1}\mathscr{H}_i^s(\varepsilon)B_{ii}(\varepsilon)R_{ii}^{-1}(\varepsilon)B_{ii}^T(\varepsilon)\sum_{r=1}^{k_i}\frac{\mu_i^r}{\mu_i^1}\mathscr{H}_i^r(\varepsilon) \\
&\quad + \sum_{s=1}^{k_i}\frac{\mu_i^s}{\mu_i^1}\mathscr{H}_i^s(\varepsilon)C_{ii}(\varepsilon)R_{zi}^{-1}(\varepsilon)C_{ii}^T(\varepsilon)\sum_{r=1}^{k_i}\frac{\mu_i^r}{\mu_i^1}\mathscr{H}_i^r(\varepsilon) \qquad (6.52)
\end{aligned}
$$

$$
\begin{aligned}
&\frac{d}{d\varepsilon}\mathscr{E}_i^r(\varepsilon) = [A_{ii}(\varepsilon) - B_{ii}(\varepsilon)R_{ii}^{-1}(\varepsilon)B_{ii}^T(\varepsilon)\sum_{s=1}^{k_i}\frac{\mu_i^s}{\mu_i^1}\mathscr{H}_i^s(\varepsilon) \\
&- C_{ii}(\varepsilon)R_{zi}^{-1}(\varepsilon)C_{ii}^T(\varepsilon)\sum_{s=1}^{k_i}\frac{\mu_i^s}{\mu_i^1}\mathscr{H}_i^s(\varepsilon)]^T\mathscr{H}_i^r(\varepsilon) + \mathscr{H}_i^r(\varepsilon)\big[A_{ii}(\varepsilon) \\
&- B_{ii}(\varepsilon)R_{ii}^{-1}(\varepsilon)B_{ii}^T(\varepsilon)\sum_{s=1}^{k_i}\frac{\mu_i^s}{\mu_i^1}\mathscr{H}_i^s(\varepsilon) - C_{ii}(\varepsilon)R_{zi}^{-1}(\varepsilon)C_{ii}^T(\varepsilon)\sum_{s=1}^{k_i}\frac{\mu_i^s}{\mu_i^1}\mathscr{H}_i^s(\varepsilon)\big] \\
&+ \sum_{v=1}^{r-1}\frac{2r!}{v!(r-v)!}\mathscr{H}_i^v(\varepsilon)G_i(\varepsilon)W_iG_i^T(\varepsilon)\mathscr{H}_i^{r-v}(\varepsilon) \\
&+ \sum_{v=1}^{r-1}\frac{2r!}{v!(r-v)!}\mathscr{H}_i^{k_i+v}(\varepsilon)G_{ic}(\varepsilon)V_cG_{ic}^T(\varepsilon)\mathscr{H}_i^{2k_i+r-v}(\varepsilon), \quad 2 \le r \le k_i \qquad (6.53)
\end{aligned}
$$

$$
\begin{aligned}
\frac{d}{d\varepsilon}\breve{\mathscr{T}}_i^1(\varepsilon) &= \big[A_{ii}(\varepsilon) - B_{ii}(\varepsilon)R_{ii}^{-1}(\varepsilon)B_{ii}^T(\varepsilon)\sum_{s=1}^{k_i}\frac{\mu_i^s}{\mu_i^1}\mathscr{H}_i^s(\varepsilon) \\
&\quad - C_{ii}(\varepsilon)R_{zi}^{-1}(\varepsilon)C_{ii}^T(\varepsilon)\sum_{s=1}^{k_i}\frac{\mu_i^s}{\mu_i^1}\mathscr{H}_i^s(\varepsilon)\big]^T\mathscr{D}_i^1(\varepsilon) \\
&\quad + \mathscr{H}_i^1(\varepsilon)\big[-B_{ii}(\varepsilon)R_{ii}^{-1}(\varepsilon)B_{ii}^T(\varepsilon)\sum_{s=1}^{k_i}\frac{\mu_i^s}{\mu_i^1}\breve{\mathscr{D}}_i^s(\varepsilon) \\
&\quad - C_{ii}(\varepsilon)R_{zi}^{-1}(\varepsilon)C_{ii}^T(\varepsilon)\sum_{s=1}^{k_i}\frac{\mu_i^s}{\mu_i^1}\breve{\mathscr{D}}_i^s(\varepsilon) + \sum_{j=1}^{N_i}B_{ij}(\varepsilon)u_{ij}^*(\varepsilon)\big]
\end{aligned}
$$

$$+\mathscr{H}_i^{k_i+1}(\varepsilon)B_{ic}(\varepsilon)u_{ic}(\varepsilon)-Q_i(\varepsilon)\zeta_i(\varepsilon)$$
$$+\sum_{s=1}^{k_i}\frac{\mu_i^s}{\mu_i^1}\mathscr{H}_i^s(\varepsilon)B_{ii}(\varepsilon)R_{ii}^{-1}(\varepsilon)B_{ii}^T(\varepsilon)\sum_{r=1}^{k_i}\frac{\mu_i^r}{\mu_i^1}\breve{\mathscr{D}}_i^r(\varepsilon)$$
$$+\sum_{s=1}^{k_i}\frac{\mu_i^s}{\mu_i^1}\mathscr{H}_i^s(\varepsilon)C_{ii}(\varepsilon)R_{zi}^{-1}(\varepsilon)C_{ii}^T(\varepsilon)\sum_{r=1}^{k_i}\frac{\mu_i^r}{\mu_i^1}\breve{\mathscr{D}}_i^r(\varepsilon) \qquad (6.54)$$

$$\frac{d}{d\varepsilon}\breve{\mathscr{D}}_i^r(\varepsilon)=\Big[A_{ii}(\varepsilon)-B_{ii}(\varepsilon)R_{ii}^{-1}(\varepsilon)B_{ii}^T(\varepsilon)\sum_{s=1}^{k_i}\frac{\mu_i^s}{\mu_i^1}\mathscr{H}_i^s(\varepsilon)$$
$$-C_{ii}(\varepsilon)R_{zi}^{-1}(\varepsilon)C_{ii}^T(\varepsilon)\sum_{s=1}^{k_i}\frac{\mu_i^s}{\mu_i^1}\mathscr{H}_i^s(\varepsilon)\Big]^T\mathscr{D}_i^r(\varepsilon)$$
$$+\mathscr{H}_i^r(\varepsilon)\Big[-B_{ii}(\varepsilon)R_{ii}^{-1}(\varepsilon)B_{ii}^T(\varepsilon)\sum_{s=1}^{k_i}\frac{\mu_i^s}{\mu_i^1}\breve{\mathscr{D}}_i^s(\varepsilon)$$
$$-C_{ii}(\varepsilon)R_{zi}^{-1}(\varepsilon)C_{ii}^T(\varepsilon)\sum_{s=1}^{k_i}\frac{\mu_i^s}{\mu_i^1}\breve{\mathscr{D}}_i^s(\varepsilon)+\sum_{j=1}^{N_i}B_{ij}(\varepsilon)u_{ij}^*(\varepsilon)\Big]$$
$$+\mathscr{H}_i^{k_i+r}(\varepsilon)B_{ic}(\varepsilon)u_{ic}(\varepsilon),\qquad 2\le r\le k_i \qquad (6.55)$$

$$\frac{d}{d\varepsilon}\mathscr{T}_i^1(\varepsilon)=\mathrm{Tr}\{\mathscr{H}_i^1(\varepsilon)G_i(\varepsilon)W_iG_i^T(\varepsilon)+\mathscr{H}_i^{3k_i+1}(\varepsilon)G_{ic}(\varepsilon)V_cG_{ic}^T(\varepsilon)\}$$
$$+2(\breve{\mathscr{D}}_i^1)^T(\varepsilon)\Big[-B_{ii}(\varepsilon)R_{ii}^{-1}(\varepsilon)B_{ii}^T(\varepsilon)\sum_{s=1}^{k_i}\frac{\mu_i^s}{\mu_i^1}\breve{\mathscr{D}}_i^s(\varepsilon)$$
$$-C_{ii}(\varepsilon)R_{zi}^{-1}(\varepsilon)C_{ii}^T(\varepsilon)\sum_{s=1}^{k_i}\frac{\mu_i^s}{\mu_i^1}\breve{\mathscr{D}}_i^s(\varepsilon)+\sum_{j=1}^{N_i}B_{ij}(\varepsilon)u_{ij}^*(\varepsilon)\Big]$$
$$+2(\breve{\mathscr{D}}_i^{k_i+1})^T(\varepsilon)B_{ic}(\varepsilon)u_{ic}(\varepsilon)+\zeta_i^T(\varepsilon)Q_i(\varepsilon)\zeta_i(\varepsilon)$$
$$+\sum_{r=1}^{k_i}\frac{\mu_i^r}{\mu_i^1}(\breve{\mathscr{D}}_i^r)^T(\varepsilon)B_{ii}(\varepsilon)R_{ii}^{-1}(\varepsilon)B_{ii}^T(\varepsilon)\sum_{s=1}^{k_i}\frac{\mu_i^s}{\mu_i^1}\breve{\mathscr{D}}_i^s(\varepsilon)$$
$$+\sum_{r=1}^{k_i}\frac{\mu_i^r}{\mu_i^1}(\breve{\mathscr{D}}_i^r)^T(\varepsilon)C_{ii}(\varepsilon)R_{zi}^{-1}(\varepsilon)C_{ii}^T(\varepsilon)\sum_{s=1}^{k_i}\frac{\mu_i^s}{\mu_i^1}\breve{\mathscr{D}}_i^s(\varepsilon) \qquad (6.56)$$

$$\frac{d}{d\varepsilon}\mathscr{T}_i^r(\varepsilon)=\mathrm{Tr}\{\mathscr{H}_i^r(\varepsilon)G_i(\varepsilon)W_iG_i^T(\varepsilon)+\mathscr{H}_i^{3k_i+r}(\varepsilon)G_{ic}(\varepsilon)V_cG_{ic}^T(\varepsilon)\}$$
$$+2(\breve{\mathscr{D}}_i^r)^T(\varepsilon)\Big[-B_{ii}(\varepsilon)R_{ii}^{-1}(\varepsilon)B_{ii}^T(\varepsilon)\sum_{s=1}^{k_i}\frac{\mu_i^s}{\mu_i^1}\breve{\mathscr{D}}_i^s(\varepsilon)$$
$$-C_{ii}(\varepsilon)R_{zi}^{-1}(\varepsilon)C_{ii}^T(\varepsilon)\sum_{s=1}^{k_i}\frac{\mu_i^s}{\mu_i^1}\breve{\mathscr{D}}_i^s(\varepsilon)+\sum_{j=1}^{N_i}B_{ij}(\varepsilon)u_{ij}^*(\varepsilon)\Big]$$
$$+2(\breve{\mathscr{D}}_i^{k_i+r})^T(\varepsilon)B_{ic}(\varepsilon)u_{ic}(\varepsilon),\quad 2\le r\le k_i \qquad (6.57)$$

will work. Furthermore, the boundary condition associated with the verification theorem requires that

$$(x_i^0)^T \sum_{r=1}^{k_i} \mu_i^r (\mathscr{H}_i^r(t_0) + \mathscr{E}_i^r(t_0)) x_i^0 + 2(x_i^0)^T \sum_{r=1}^{k_i} \mu_i^r (\breve{\mathscr{D}}_i^r(t_0) + \breve{\mathscr{T}}_i^r(t_0))$$
$$+ \sum_{r=1}^{k_i} \mu_i^r (\mathscr{D}_i^r(t_0) + \mathscr{T}_i^r(t_0))$$
$$= (x_i^0)^T \sum_{r=1}^{k_i} \mu_i^r \mathscr{H}_i^r(t_0) x_i^0 + 2(x_i^0)^T \sum_{r=1}^{k_i} \mu_i^r \breve{\mathscr{D}}_i^r(t_0) + \sum_{r=1}^{k_i} \mu_i^r \mathscr{D}_i^r(t_0) .$$

Thus, matching the boundary condition yields the initial value conditions $\mathscr{E}_i^r(t_0) = 0$, $\breve{\mathscr{T}}_i^r(t_0) = 0$ and $\mathscr{T}_i^r(t_0) = 0$ for the forward-in-time differential equations (6.52)-(6.57).

Applying the 4-tuple $(K_{x_i}, K_{z_i}, p_{x_i}, p_{z_i})$ in (6.47)-(6.50) that is defining the person-by-person equilibrium for each agent i and $i \in \overline{I}$ along the solution trajectories of the backward-in-time differential equations (6.33)-(6.35), these equations become the backward-in-time Riccati-type differential equations

$$\frac{d}{d\varepsilon}\mathscr{H}_i^1(\varepsilon) = -\Big[A_{ii}(\varepsilon) - B_{ii}(\varepsilon)R_{ii}^{-1}(\varepsilon)B_{ii}^T(\varepsilon)\sum_{s=1}^{k_i}\frac{\mu_i^s}{\mu_i^1}\mathscr{H}_i^s(\varepsilon)$$
$$- C_{ii}(\varepsilon)R_{zi}^{-1}(\varepsilon)C_{ii}^T(\varepsilon)\sum_{s=1}^{k_i}\frac{\mu_i^s}{\mu_i^1}\mathscr{H}_i^s(\varepsilon)\Big]^T \mathscr{H}_i^1(\varepsilon)$$
$$- \mathscr{H}_i^1(\varepsilon)\Big[A_{ii}(\varepsilon) - B_{ii}(\varepsilon)R_{ii}^{-1}(\varepsilon)B_{ii}^T(\varepsilon)\sum_{s=1}^{k_i}\frac{\mu_i^s}{\mu_i^1}\mathscr{H}_i^s(\varepsilon)$$
$$- C_{ii}(\varepsilon)R_{zi}^{-1}(\varepsilon)C_{ii}^T(\varepsilon)\sum_{s=1}^{k_i}\frac{\mu_i^s}{\mu_i^1}\mathscr{H}_i^s(\varepsilon)\Big] - Q_{ii}(\varepsilon) - Q_i(\varepsilon)$$
$$- \sum_{s=1}^{k_i}\frac{\mu_i^s}{\mu_i^1}\mathscr{H}_i^s(\varepsilon)B_{ii}(\varepsilon)R_{ii}^{-1}(\varepsilon)B_{ii}^T(\varepsilon)\sum_{r=1}^{k_i}\frac{\mu_i^r}{\mu_i^1}\mathscr{H}_i^r(\varepsilon)$$
$$- \sum_{s=1}^{k_i}\frac{\mu_i^s}{\mu_i^1}\mathscr{H}_i^s(\varepsilon)C_{ii}(\varepsilon)R_{zi}^{-1}(\varepsilon)C_{ii}^T(\varepsilon)\sum_{r=1}^{k_i}\frac{\mu_i^r}{\mu_i^1}\mathscr{H}_i^r(\varepsilon) \quad (6.58)$$

$$
\begin{aligned}
\frac{d}{d\varepsilon}\mathscr{H}_i^r(\varepsilon) = &-\Big[A_{ii}(\varepsilon) - B_{ii}(\varepsilon)R_{ii}^{-1}(\varepsilon)B_{ii}^T(\varepsilon)\sum_{s=1}^{k_i}\frac{\mu_i^s}{\mu_i^1}\mathscr{H}_i^s(\varepsilon) \\
&- C_{ii}(\varepsilon)R_{zi}^{-1}(\varepsilon)C_{ii}^T(\varepsilon)\sum_{s=1}^{k_i}\frac{\mu_i^s}{\mu_i^1}\mathscr{H}_i^s(\varepsilon)\Big]^T\mathscr{H}_i^r(\varepsilon) - \mathscr{H}_i^r(\varepsilon)\Big[A_{ii}(\varepsilon) \\
&- B_{ii}(\varepsilon)R_{ii}^{-1}(\varepsilon)B_{ii}^T(\varepsilon)\sum_{s=1}^{k_i}\frac{\mu_i^s}{\mu_i^1}\mathscr{H}_i^s(\varepsilon) - C_{ii}(\varepsilon)R_{zi}^{-1}(\varepsilon)C_{ii}^T(\varepsilon)\sum_{s=1}^{k_i}\frac{\mu_i^s}{\mu_i^1}\mathscr{H}_i^s(\varepsilon)\Big] \\
&- \sum_{v=1}^{r-1}\frac{2r!}{v!(r-v)!}\mathscr{H}_i^v(\varepsilon)G_i(\varepsilon)W_iG_i^T(\varepsilon)\mathscr{H}_i^{r-v}(\varepsilon) \\
&- \sum_{v=1}^{r-1}\frac{2r!}{v!(r-v)!}\mathscr{H}_i^{k_i+v}(\varepsilon)G_{ic}(\varepsilon)V_cG_{ic}^T(\varepsilon)\mathscr{H}_i^{2k_i+r-v}(\varepsilon), \quad 2 \le r \le k_i \qquad (6.59)
\end{aligned}
$$

$$
\begin{aligned}
\frac{d}{d\varepsilon}\breve{\mathscr{D}}_i^1(\varepsilon) = &-\Big[A_{ii}(\varepsilon) - B_{ii}(\varepsilon)R_{ii}^{-1}(\varepsilon)B_{ii}^T(\varepsilon)\sum_{s=1}^{k_i}\frac{\mu_i^s}{\mu_i^1}\mathscr{H}_i^s(\varepsilon) \\
&- C_{ii}(\varepsilon)R_{zi}^{-1}(\varepsilon)C_{ii}^T(\varepsilon)\sum_{s=1}^{k_i}\frac{\mu_i^s}{\mu_i^1}\mathscr{H}_i^s(\varepsilon)\Big]^T\mathscr{D}_i^1(\varepsilon) \\
&- \mathscr{H}_i^1(\varepsilon)\Big[-B_{ii}(\varepsilon)R_{ii}^{-1}(\varepsilon)B_{ii}^T(\varepsilon)\sum_{s=1}^{k_i}\frac{\mu_i^s}{\mu_i^1}\breve{\mathscr{D}}_i^s(\varepsilon) \\
&- C_{ii}(\varepsilon)R_{zi}^{-1}(\varepsilon)C_{ii}^T(\varepsilon)\sum_{s=1}^{k_i}\frac{\mu_i^s}{\mu_i^1}\breve{\mathscr{D}}_i^s(\varepsilon) + \sum_{j=1}^{N_i}B_{ij}(\varepsilon)u_{ij}^*(\varepsilon)\Big] \\
&- \mathscr{H}_i^{k_i+1}(\varepsilon)B_{ic}(\varepsilon)u_{ic}(\varepsilon) + Q_i(\varepsilon)\zeta_i(\varepsilon) \\
&- \sum_{s=1}^{k_i}\frac{\mu_i^s}{\mu_i^1}\mathscr{H}_i^s(\varepsilon)B_{ii}(\varepsilon)R_{ii}^{-1}(\varepsilon)B_{ii}^T(\varepsilon)\sum_{r=1}^{k_i}\frac{\mu_i^r}{\mu_i^1}\breve{\mathscr{D}}_i^r(\varepsilon) \\
&- \sum_{s=1}^{k_i}\frac{\mu_i^s}{\mu_i^1}\mathscr{H}_i^s(\varepsilon)C_{ii}(\varepsilon)R_{zi}^{-1}(\varepsilon)C_{ii}^T(\varepsilon)\sum_{r=1}^{k_i}\frac{\mu_i^r}{\mu_i^1}\breve{\mathscr{D}}_i^r(\varepsilon) \qquad (6.60)
\end{aligned}
$$

$$\frac{d}{d\varepsilon}\breve{\mathscr{D}}_i^r(\varepsilon) = -\big[A_{ii}(\varepsilon) - B_{ii}(\varepsilon)R_{ii}^{-1}(\varepsilon)B_{ii}^T(\varepsilon)\sum_{s=1}^{k_i}\frac{\mu_i^s}{\mu_i^1}\mathscr{H}_i^s(\varepsilon)$$
$$- C_{ii}(\varepsilon)R_{zi}^{-1}(\varepsilon)C_{ii}^T(\varepsilon)\sum_{s=1}^{k_i}\frac{\mu_i^s}{\mu_i^1}\mathscr{H}_i^s(\varepsilon)\big]^T\mathscr{D}_i^r(\varepsilon)$$
$$- \mathscr{H}_i^r(\varepsilon)\big[-B_{ii}(\varepsilon)R_{ii}^{-1}(\varepsilon)B_{ii}^T(\varepsilon)\sum_{s=1}^{k_i}\frac{\mu_i^s}{\mu_i^1}\breve{\mathscr{D}}_i^s(\varepsilon)$$
$$- C_{ii}(\varepsilon)R_{zi}^{-1}(\varepsilon)C_{ii}^T(\varepsilon)\sum_{s=1}^{k_i}\frac{\mu_i^s}{\mu_i^1}\breve{\mathscr{D}}_i^s(\varepsilon) + \sum_{j=1}^{N_i}B_{ij}(\varepsilon)u_{ij}^*(\varepsilon)\big]$$
$$- \mathscr{H}_i^{k_i+r}(\varepsilon)B_{ic}(\varepsilon)u_{ic}(\varepsilon), \qquad 2 \le r \le k_i \tag{6.61}$$

$$\frac{d}{d\varepsilon}\mathscr{D}_i^1(\varepsilon) = -\mathrm{Tr}\{\mathscr{H}_i^1(\varepsilon)G_i(\varepsilon)W_iG_i^T(\varepsilon) + \mathscr{H}_i^{3k_i+1}(\varepsilon)G_{ic}(\varepsilon)V_cG_{ic}^T(\varepsilon)\}$$
$$- 2(\breve{\mathscr{D}}_i^1)^T(\varepsilon)\big[-B_{ii}(\varepsilon)R_{ii}^{-1}(\varepsilon)B_{ii}^T(\varepsilon)\sum_{s=1}^{k_i}\frac{\mu_i^s}{\mu_i^1}\breve{\mathscr{D}}_i^s(\varepsilon)$$
$$- C_{ii}(\varepsilon)R_{zi}^{-1}(\varepsilon)C_{ii}^T(\varepsilon)\sum_{s=1}^{k_i}\frac{\mu_i^s}{\mu_i^1}\breve{\mathscr{D}}_i^s(\varepsilon) + \sum_{j=1}^{N_i}B_{ij}(\varepsilon)u_{ij}^*(\varepsilon)\big]$$
$$- 2(\breve{\mathscr{D}}_i^{k_i+1})^T(\varepsilon)B_{ic}(\varepsilon)u_{ic}(\varepsilon) - \zeta_i^T(\varepsilon)Q_i(\varepsilon)\zeta_i(\varepsilon)$$
$$- \sum_{r=1}^{k_i}\frac{\mu_i^r}{\mu_i^1}(\breve{\mathscr{D}}_i^r)^T(\varepsilon)B_{ii}(\varepsilon)R_{ii}^{-1}(\varepsilon)B_{ii}^T(\varepsilon)\sum_{s=1}^{k_i}\frac{\mu_i^s}{\mu_i^1}\breve{\mathscr{D}}_i^s(\varepsilon)$$
$$- \sum_{r=1}^{k_i}\frac{\mu_i^r}{\mu_i^1}(\breve{\mathscr{D}}_i^r)^T(\varepsilon)C_{ii}(\varepsilon)R_{zi}^{-1}(\varepsilon)C_{ii}^T(\varepsilon)\sum_{s=1}^{k_i}\frac{\mu_i^s}{\mu_i^1}\breve{\mathscr{D}}_i^s(\varepsilon) \tag{6.62}$$

$$\frac{d}{d\varepsilon}\mathscr{D}_i^r(\varepsilon) = -\mathrm{Tr}\{\mathscr{H}_i^r(\varepsilon)G_i(\varepsilon)W_iG_i^T(\varepsilon) + \mathscr{H}_i^{3k_i+r}(\varepsilon)G_{ic}(\varepsilon)V_cG_{ic}^T(\varepsilon)\}$$
$$- 2(\breve{\mathscr{D}}_i^r)^T(\varepsilon)\big[-B_{ii}(\varepsilon)R_{ii}^{-1}(\varepsilon)B_{ii}^T(\varepsilon)\sum_{s=1}^{k_i}\frac{\mu_i^s}{\mu_i^1}\breve{\mathscr{D}}_i^s(\varepsilon)$$
$$- C_{ii}(\varepsilon)R_{zi}^{-1}(\varepsilon)C_{ii}^T(\varepsilon)\sum_{s=1}^{k_i}\frac{\mu_i^s}{\mu_i^1}\breve{\mathscr{D}}_i^s(\varepsilon) + \sum_{j=1}^{N_i}B_{ij}(\varepsilon)u_{ij}^*(\varepsilon)\big]$$
$$- 2(\breve{\mathscr{D}}_i^{k_i+r})^T(\varepsilon)B_{ic}(\varepsilon)u_{ic}(\varepsilon), \quad 2 \le r \le k_i \tag{6.63}$$

where the terminal-value conditions $\mathscr{H}_i^1(t_f) = Q_i^f$, $\mathscr{H}_i^r(t_f) = 0$ for $2 \le r \le k_i$; $\breve{\mathscr{D}}_i^r(t_f) = -Q_i^f\zeta_i(t_f)$, $\breve{\mathscr{D}}_i^r(t_f) = 0$ for $2 \le r \le k_i$; and $\mathscr{D}_i^r(t_f) = \zeta_i(t_f)Q_i^f\zeta_i(t_f)$, $\mathscr{D}_i^r(t_f)$ for $2 \le r \le k_i$. Thus, whenever the coupled backward-in-time differential equations (6.58)-(6.63) admit the matrix-valued solutions $\{\mathscr{H}_i^r(\cdot)\}_{r=1}^{k_i}$, vector-valued solutions $\{\breve{\mathscr{D}}_i^r(\cdot)\}_{r=1}^{k_i}$ and scalar-valued solutions $\{\mathscr{D}_i^r(\cdot)\}_{r=1}^{k_i}$, then the existence of the matrix-valued solutions $\{\mathscr{E}_i^r(\cdot)\}_{r=1}^{k_i}$, vector-valued solutions $\{\breve{\mathscr{T}}_i^r(\cdot)\}_{r=1}^{k_i}$ and scalar-valued solutions $\{\mathscr{T}_i^r(\cdot)\}_{r=1}^{k_i}$ satisfying the coupled forward-in-time differential equations (6.52)-(6.57) are assured. By comparing the time-forward differential equations (6.52)-(6.57) to those of time-backward differential equations (6.58)-(6.63), one may recognize that these sets of differential equations are related to one another by

$$\frac{d}{d\varepsilon}\mathscr{E}_i^r(\varepsilon) = -\frac{d}{d\varepsilon}\mathscr{H}_i^r(\varepsilon); \quad \frac{d}{d\varepsilon}\breve{\mathscr{T}}_i^r(\varepsilon) = -\frac{d}{d\varepsilon}\breve{\mathscr{D}}_i^r(\varepsilon); \quad \frac{d}{d\varepsilon}\mathscr{T}_i^r(\varepsilon) = -\frac{d}{d\varepsilon}\mathscr{D}_i^r(\varepsilon).$$

Enforcing the initial-value conditions of $\mathscr{E}_i^r(t_0) = 0$, $\breve{\mathscr{T}}_i^r(t_0) = 0$ and $\mathscr{T}_i^r(t_0) = 0$ uniquely implies the following results

$$\mathscr{E}_i^r(\varepsilon) = \mathscr{H}_i^r(t_0) - \mathscr{H}_i^r(\varepsilon); \quad \breve{\mathscr{T}}_i^r(\varepsilon) = \breve{\mathscr{D}}_i^r(t_0) - \breve{\mathscr{D}}_i^r(\varepsilon); \quad \mathscr{T}_i^r(\varepsilon) = \mathscr{D}_i^r(t_0) - \mathscr{D}_i^r(\varepsilon)$$

for all $\varepsilon \in [t_0, t_f]$ and yields a value function

$$\mathscr{W}_i(\varepsilon, \mathscr{Y}_i, \breve{\mathscr{Z}}_i, \mathscr{Z}_i) = (x_i^0)^T \sum_{r=1}^{k_i} \mu_i^r \mathscr{H}_i^r(t_0) x_i^0 + 2(x_i^0)^T \sum_{r=1}^{k_i} \mu_i^r \breve{\mathscr{D}}_i^r(t_0) + \sum_{r=1}^{k_i} \mu_i^r \mathscr{D}_i^r(t_0)$$

for which the sufficient condition (6.42) of the verification theorem is satisfied. Therefore, the extremal person-by-person equilibrium policy (6.47)-(6.50) minimizing (6.36) become optimal

$$K_{x_i}^*(\varepsilon) = -R_{ii}^{-1}(\varepsilon) B_{ii}^T(\varepsilon) \sum_{r=1}^{k_i} \hat{\mu}_i^r \mathscr{H}_i^{*r}(\varepsilon), \tag{6.64}$$

$$K_{z_i}^*(\varepsilon) = -R_{zi}^{-1}(\varepsilon) C_{ii}^T(\varepsilon) \sum_{r=1}^{k_i} \hat{\mu}_i^r \mathscr{H}_i^{*r}(\varepsilon), \tag{6.65}$$

$$p_{x_i}^*(\varepsilon) = -R_{ii}^{-1}(\varepsilon) B_{ii}^T(\varepsilon) \sum_{r=1}^{k_i} \hat{\mu}_i^r \breve{\mathscr{D}}_i^{*r}(\varepsilon), \tag{6.66}$$

$$p_{z_i}^*(\varepsilon) = -R_{zi}^{-1}(\varepsilon) C_{ii}^T(\varepsilon) \sum_{r=1}^{k_i} \hat{\mu}_i^r \breve{\mathscr{D}}_i^{*r}(\varepsilon), \tag{6.67}$$

whereby $\hat{\mu}_i^r = \frac{\mu_i^r}{\mu_i^1}$.

The goals in this research investigation have been methodological. A noncooperative game-theoretic methodology for coordination control of distributed stochastic systems is successfully sought for theory building in contexts in which signaling effects are issued by a coordinator and distributed person-by-person equilibrium strategies by autonomous agents i and $i \in \overline{I}$ are placed toward performance robustness. At this point, it makes sense to integrate all of the contending results into the following unified theorem.

Theorem 5 *Person-by-Person Equilibrium Strategies.*
Consider a distributed stochastic system governed by (6.3)-(6.7) whose pairs (A_{ii}, B_{ii}) and (A_{ii}, C_{ii}) are uniformly stabilizable on $[t_0, t_f]$. A N-tuple $\{(u_1^*, z_1^*), \ldots, (u_N^*, z_N^*)\}$ of control policies constitutes a feedback Nash equilibrium for the class of distributed stochastic system considered here. Furthermore, 2-tuple (u_i^*, z_i^*) imposing a person-by-person equilibrium strategy for the corresponding agent i and $i \in \bar{I}$ is implemented forwardly in time by

$$u_i^*(t) = K_{x_i}^*(t)x_i(t) + p_{x_i}^*(t) \tag{6.68}$$

$$z_i^*(t) = K_{z_i}^*(t)x_i(t) + p_{z_i}^*(t), \quad t = t_f + t_0 - \varepsilon, \quad \varepsilon \in [t_0, t_f], \tag{6.69}$$

which strives to optimize the risk-value aware performance index (6.36) composed by a preferential set of mathematical statistics of the chi-squared cost random variable (6.7). The construction of the person-by-person equilibrium for each agent i is determined backwardly in time; e.g.,

$$K_{x_i}^*(\varepsilon) = -R_{ii}^{-1}(\varepsilon)B_{ii}^T(\varepsilon)\sum_{r=1}^{k_i} \hat{\mu}_i{}^r \mathscr{H}_i^{*r}(\varepsilon), \tag{6.70}$$

$$K_{z_i}^*(\varepsilon) = -R_{zi}^{-1}(\varepsilon)C_{ii}^T(\varepsilon)\sum_{r=1}^{k_i} \hat{\mu}_i{}^r \mathscr{H}_i^{*r}(\varepsilon), \tag{6.71}$$

$$p_{x_i}^*(\varepsilon) = -R_{ii}^{-1}(\varepsilon)B_{ii}^T(\varepsilon)\sum_{r=1}^{k_i} \hat{\mu}_i{}^r \breve{\mathscr{D}}_i^{*r}(\varepsilon), \tag{6.72}$$

$$p_{z_i}^*(\varepsilon) = -R_{zi}^{-1}(\varepsilon)C_{ii}^T(\varepsilon)\sum_{r=1}^{k_i} \hat{\mu}_i{}^r \breve{\mathscr{D}}_i^{*r}(\varepsilon), \tag{6.73}$$

wherein the normalized preferences $\hat{\mu}_i^r \triangleq \mu_i^r/\mu_i^1$'s are mutually chosen by each incumbent agent i for risk-averse coordinations toward co-design of individual performance robustness. The optimal set of supporting solutions $\{\mathscr{H}_{i*}^r(\varepsilon)\}_{r=1}^{k_i}$, $\{\mathscr{H}_{i*}^{k_i+r}(\varepsilon)\}_{r=1}^{k_i}$, $\{\mathscr{H}_{i*}^{2k_i+r}(\varepsilon)\}_{r=1}^{k_i}$, $\{\mathscr{H}_{i*}^{3k_i+r}(\varepsilon)\}_{r=1}^{k_i}$ and $\{\breve{\mathscr{D}}_{i*}^r(\varepsilon)\}_{r=1}^{k_i}$ satisfy the time-backward and matrix-valued differential equations

$$\begin{aligned}\frac{d}{d\varepsilon}\mathscr{H}_{i*}^1(\varepsilon) = &-[A_{ii}(\varepsilon) + B_{ii}(\varepsilon)K_{x_i}^*(\varepsilon) + C_{ii}(\varepsilon)K_{z_i}^*(\varepsilon)]^T \mathscr{H}_{i*}^1(\varepsilon)\\ &- \mathscr{H}_{i*}^1(\varepsilon)[A_{ii}(\varepsilon) + B_{ii}(\varepsilon)K_{x_i}^*(\varepsilon) + C_{ii}(\varepsilon)K_{z_i}^*(\varepsilon)]\\ &- Q_{ii}(\varepsilon) - Q_i(\varepsilon) - K_{x_i}^{*T}(\varepsilon)R_{ii}(\varepsilon)K_{x_i}^*(\varepsilon) - K_{z_i}^{*T}(\varepsilon)R_{zi}(\varepsilon)K_{z_i}^*(\varepsilon)\end{aligned} \tag{6.74}$$

$$
\begin{aligned}
\frac{d}{d\varepsilon}\mathscr{H}_{i*}^{r}(\varepsilon) &= -[A_{ii}(\varepsilon)+B_{ii}(\varepsilon)K_{x_i}^{*}(\varepsilon)+C_{ii}(\varepsilon)K_{z_i}^{*}(\varepsilon)]^T\mathscr{H}_{i*}^{r}(\varepsilon)\\
&\quad -\mathscr{H}_{i*}^{r}(\varepsilon)[A_{ii}(\varepsilon)+B_{ii}(\varepsilon)K_{x_i}^{*}(\varepsilon)+C_{ii}(\varepsilon)K_{z_i}^{*}(\varepsilon)]\\
&\quad -\sum_{v=1}^{r-1}\frac{2r!}{v!(r-v)!}\mathscr{H}_{i*}^{v}(\varepsilon)G_i(\varepsilon)W_iG_i^T(\varepsilon)\mathscr{H}_{i*}^{r-v}(\varepsilon)\\
&\quad -\sum_{v=1}^{r-1}\frac{2r!}{v!(r-v)!}\mathscr{H}_{i*}^{k_i+v}(\varepsilon)G_{ic}(\varepsilon)V_cG_{ic}^T(\varepsilon)\mathscr{H}_{i*}^{2k_i+r-v}(\varepsilon), \quad 2\le r\le k_i
\end{aligned}
\tag{6.75}
$$

$$
\begin{aligned}
\frac{d}{d\varepsilon}\mathscr{H}_{i*}^{k_i+1}(\varepsilon) &= -[A_{ii}(\varepsilon)+B_{ii}(\varepsilon)K_{x_i}^{*}(\varepsilon)+C_{ii}(s)K_{z_i}^{*}(\varepsilon)]^T\mathscr{H}_{i*}^{k_i+1}(\varepsilon)\\
&\quad -\mathscr{H}_{i*}^{k_i+1}(\varepsilon)A_{ic}(\varepsilon)+2K_{z_i}^{*T}(\varepsilon)R_{zi}(\varepsilon)
\end{aligned}
\tag{6.76}
$$

$$
\begin{aligned}
\frac{d}{d\varepsilon}\mathscr{H}_{i*}^{k_i+r}(\varepsilon) &= -[A_{ii}(\varepsilon)+B_{ii}(\varepsilon)K_{x_i}^{*}(\varepsilon)+C_{ii}(\varepsilon)K_{z_i}^{*}(\varepsilon)]^T\mathscr{H}_{i*}^{k_i+r}(\varepsilon)\\
&\quad -\mathscr{H}_{i*}^{k_i+r}(\varepsilon)A_{ic}(\varepsilon)-\sum_{v=1}^{r-1}\frac{2r!}{v!(r-v)!}\mathscr{H}_{i*}^{v}(\varepsilon)G_i(\varepsilon)W_iG_i^T(\varepsilon)\mathscr{H}_{i*}^{k_i+r-v}(\varepsilon)\\
&\quad -\sum_{v=1}^{r-1}\frac{2r!}{v!(r-v)!}\mathscr{H}_{i*}^{k_i+v}(\varepsilon)G_{ic}(\varepsilon)V_cG_{ic}^T(\varepsilon)\mathscr{H}_{i*}^{3k_i+r-v}(\varepsilon), \quad 2\le r\le k_i
\end{aligned}
\tag{6.77}
$$

$$
\begin{aligned}
\frac{d}{d\varepsilon}\mathscr{H}_{i*}^{2k_i+1}(\varepsilon) &= -A_{ic}^T(\varepsilon)\mathscr{H}_{i*}^{2k_i+1}(\varepsilon)\\
&\quad -\mathscr{H}_{i*}^{2k_i+1}(\varepsilon)[A_{ii}(\varepsilon)+B_{ii}(\varepsilon)K_{x_i}^{*}(\varepsilon)+C_{ii}(\varepsilon)K_{z_i}^{*}(\varepsilon)]
\end{aligned}
\tag{6.78}
$$

$$
\begin{aligned}
\frac{d}{d\varepsilon}\mathscr{H}_{i*}^{2k_i+r}(\varepsilon) &= -\mathscr{H}_{i*}^{2k_i+r}(\varepsilon)[A_{ii}(\varepsilon)+B_{ii}(\varepsilon)K_{x_i}^{*}(\varepsilon)+C_{ii}(\varepsilon)K_{z_i}^{*}(\varepsilon)]\\
&\quad -A_{ic}^T(\varepsilon)\mathscr{H}_{i*}^{2k_i+r}(\varepsilon)-\sum_{v=1}^{r-1}\frac{2r!}{v!(r-v)!}\mathscr{H}_{i*}^{2k_i+v}(\varepsilon)G_i(\varepsilon)W_iG_i^T(\varepsilon)\mathscr{H}_{i*}^{r-v}(\varepsilon)\\
&\quad -\sum_{v=1}^{r-1}\frac{2r!}{v!(r-v)!}\mathscr{H}_{i*}^{3k_i+v}(\varepsilon)G_{ic}(\varepsilon)V_cG_{ic}^T(\varepsilon)\mathscr{H}_{i*}^{2k_i+r-v}(\varepsilon)], \quad 2\le r\le k_i
\end{aligned}
\tag{6.79}
$$

$$
\frac{d}{d\varepsilon}\mathscr{H}_{i*}^{3k_i+1}(\varepsilon) = -A_{ic}^T(\varepsilon)\mathscr{H}_{i*}^{3k_i+1}(\varepsilon)-\mathscr{H}_{i*}^{3k_i+1}(\varepsilon)A_{ic}(\varepsilon)-R_{zi}(\varepsilon)
\tag{6.80}
$$

$$\frac{d}{d\varepsilon}\mathscr{H}_{i*}^{3k_i+r}(\varepsilon) = -A_{ic}^T(\varepsilon)\mathscr{H}_{i*}^{3k_i+r}(\varepsilon) - \mathscr{H}_{i*}^{3k_i+r}(\varepsilon)A_{ic}(\varepsilon) - R_{zi}(\varepsilon)$$
$$- \sum_{v=1}^{r-1}\frac{2r!}{v!(r-v)!}\mathscr{H}_{i*}^{2k_i+v}(\varepsilon)G_i(\varepsilon)W_iG_i^T(\varepsilon)\mathscr{H}_{i*}^{k_i+r-v}(\varepsilon)$$
$$- \sum_{v=1}^{r-1}\frac{2r!}{v!(r-v)!}\mathscr{H}_{i*}^{3k_i+v}(\varepsilon)G_{ic}(\varepsilon)V_cG_{ic}^T(\varepsilon)\mathscr{H}_{i*}^{3k_i+r-v}(\varepsilon), \quad 2 \le r \le k_i, \qquad (6.81)$$

in addition with the backward-in-time and vector-valued differential equations

$$\frac{d}{d\varepsilon}\breve{\mathscr{D}}_{i*}^1(\varepsilon) = -[A_{ii}(\varepsilon) + B_{ii}(\varepsilon)K_{x_i}^*(\varepsilon) + C_{ii}(s)K_{z_i}^*(\varepsilon)]^T\breve{\mathscr{D}}_{i*}^1(\varepsilon)$$
$$- \mathscr{H}_{i*}^1(\varepsilon)[B_{ii}(\varepsilon)p_{x_i}^*(\varepsilon) + C_{ii}(\varepsilon)p_{z_i}^*(\varepsilon) + \sum_{j=1}^{N_i}B_{ij}(\varepsilon)u_{ij}^*(\varepsilon)] + Q_i(\varepsilon)\zeta_i(\varepsilon)$$
$$- \mathscr{H}_{i*}^{k_i+1}(\varepsilon)B_{ic}(\varepsilon)u_{ic}(\varepsilon) - K_{x_i}^{*T}(\varepsilon)R_{ii}(\varepsilon)p_{x_i}^*(\varepsilon) - K_{z_i}^{*T}R_{zi}(\varepsilon)p_{z_i}^*(\varepsilon) \qquad (6.82)$$

$$\frac{d}{d\varepsilon}\breve{\mathscr{D}}_{i*}^r(\varepsilon) = -[A_{ii}(\varepsilon) + B_{ii}(\varepsilon)K_{x_i}^*(\varepsilon) + C_{ii}(s)K_{z_i}^*(\varepsilon)]^T\breve{\mathscr{D}}_{i*}^r(\varepsilon)$$
$$- \mathscr{H}_{i*}^r(\varepsilon)[B_{ii}(\varepsilon)p_{x_i}^*(\varepsilon) + C_{ii}(\varepsilon)p_{z_i}^*(\varepsilon) + \sum_{j=1}^{N_i}B_{ij}(\varepsilon)u_{ij}^*(\varepsilon)]$$
$$- \mathscr{H}_{i*}^{k_i+r}(\varepsilon)B_{ic}(\varepsilon)u_{ic}(\varepsilon), \qquad 2 \le r \le k_i \qquad (6.83)$$

whereby the terminal-value conditions $\mathscr{H}_{i*}^1(t_f) = Q_i^f$, $\mathscr{H}_{i*}^r(t_f) = 0$ *for* $2 \le r \le k_i$; $\mathscr{H}_{i*}^{k_i+r}(t_f) = 0$ *for* $1 \le r \le k_i$; $\mathscr{H}_{i*}^{2k_i+r}(t_f) = 0$ *for* $1 \le r \le k_i$; $\mathscr{H}_{i*}^{3k_i+r}(t_f) = 0$ *for* $1 \le r \le k_i$; *and* $\breve{\mathscr{D}}_{i*}^1(t_f) = -Q_i^f\zeta_i(t_f)$, $\breve{\mathscr{D}}_{i*}^r(t_f) = 0$ *for* $2 \le r \le k_i$.

6.5 Conclusions

The noncooperative game-theoretic framework presented here offers a contribution to coordination control science's existing portfolio of methodologies. This portfolio contains a coordinator which directs two or more interconnected stochastic systems. Thinking about risk-averse attitudes toward performance uncertainty and distribution suggests new ideas for extending existing theories of distributed control and multiperson decision analysis. In this sense, the present research article offers a theoretic lens and a novel approach that direct attention towards mathematical statistics of the chi-squared random performance measures concerned by incumbent agents of the class of distributed stochastic systems herein and thus provide new insights into complex dynamics of performance robustness and reliability. To account for mutual influence from immediate neighbors that give rise to interaction complexity such as potential noncooperation, each incumbent system or self-directed agent autonomously focuses on the search for a person-by-person equilibrium which is in turn locally supported by perfect state observations. Further discussions show that the person-by-person equilibrium is equivalent to the concept of feedback Nash strategy. Research issues discussed include adjusting the risk-averse attitudes against which performance statistics are measured is now termed as risk-value aware performance indices. The process of adjustment for performance risk aversion imposes some computational requirements as needed by the construction of the states of the person-by-person equilibrium.

Future work will be another attainments of distributed control with explicit communications and partial information patterns, wherein research issues are: (i) when is periodic communication the optimal communication policy for the considered cost function? (ii) in what control and decision architectures, are partial local state information shared? and (iii) how to distribute global objectives of distributed and/or hierarchical stochastic systems over constituent systems and/or different layers?

References

1. Friedman, J.W.: Game theory with applications to economics, 2nd edn. Oxford University Press, New York (1990)
2. Fudenberg, D., Levine, D.K.: The theory of learning in games. MIT Press, Cambridge (1998)
3. Huang, M.: Large-population LQG games involving a major player: the Nash certainty equivalence principle. SIAM Journal of Control Optimization 48(5), 3318–3353 (2010)
4. Pham, K.D.: Non-cooperative outcomes for stochastic nash games: decision strategies towards multi-attribute performance robustness. In: The 17th World Congress International Federation on Automatic Control, pp. 11750–11756 (2008)
5. Pham, K.D.: New results in stochastic cooperative games: strategic coordination for multi-resolution performance robustness. In: Hirsch, M.J., Pardalos, P.M., Murphey, R., Grundel, D. (eds.) Optimization and Cooperative Control Strategies. LNCIS, vol. 381, pp. 257–285. Springer, Heidelberg (2008)
6. Pham, K.D., Liberty, S.R., Jin, G.: Multi-cumulant and Pareto solutions for tactics change prediction and performance analysis in stochastic multi-team non-cooperative games. In: Won, C.-H., Schrader, C., Michel, A.N. (eds.) Advances in Statistical Control, Algebraic Systems Theory and Dynamic Systems Characteristics. Systems & Control: Foundations and Applications, pp. 65–97. Birkhauser, Boston (2008)
7. Pham, K.D.: Performance-reliability aided decision making in multiperson quadratic decision games against jamming and estimation confrontations. Journal of Optimization Theory and Applications 149(3), 559–629 (2011)
8. Brockett, R.W.: Finite dimensional linear sytems. Wiley, New York (1970)
9. Jacobson, D.H.: Optimal stochastic linear systems with exponential performance criteria and their relation to deterministic games. IEEE Transactions on Automatic Control 18, 124–131 (1973)
10. Whittle, P.: Risk sensitive optimal control. John Wiley & Sons, New York (1990)
11. Fleming, W.H., Rishel, R.W.: Deterministic and stochastic optimal control. Springer, New York (1975)

Chapter 7
Weighted and Constrained Consensus Control with Performance Optimization for Robust Distributed UAV Deployments with Dynamic Information Networks*

Le Yi Wang and George Yin

Abstract. A team of unmanned aerial vehicles (UAVs) in surveillance operations aims to achieve fast deployments, robustness against uncertainties and adversaries, and adaptability when the team expands or reduces. All these must be achieved under time-varying and local communication connections. This paper introduces a new framework for UAV control based on the emerging consensus control for networked systems. Due to unique features of UAV tasks, the consensus control problem becomes weighted and constrained, beyond the typical consensus formulation. Using only neighborhood communications among UAV members, the consensus control achieves global desired deployment. Algorithms are introduced and their convergence properties are established. It is shown that the algorithms achieve asymptotically the Cramér-Rao lower bound, and hence is asymptotically optimal among all algorithms. Examples and case studies demonstrate convergence, robustness, and scalability of the algorithms.

Keywords: UAV control, consensus control, team coordination, networked systems, stochastic approximation algorithm.

Le Yi Wang
Department of Electrical and Computer Engineering, Wayne State University, Detroit, MI 48202
e-mail: `lywang@ece.eng.wayne.edu`

George Yin
Department of Mathematics, Wayne State University, Detroit, MI 48202
e-mail: `gyin@math.wayne.edu`

* This research was supported in part by the Air Force Office of Scientific Research under FA9550-10-1-0210.

F. Fahroo et al. (Eds.): *Recent Adv. in Res. on Unmanned Aerial Veh.*, LNCIS 444, pp. 181–205.
DOI: 10.1007/978-3-642-37694-8_7

7.1 Introduction

This study is motivated by the increasing demands of more advanced technology for information processing, control, coordination, and dynamic reconfiguration of networked unmanned aerial vehicles (UAVs). A cluster of UAVs for a coordinated task such as a surveillance mission forms a networked system; see Figure 7.1. Each subsystem is represented by a node which is a local dynamic system itself, and communication and connections of subsystems are represented by a network topology. The networked system aims to accomplish a joint mission, in the presence of uncertainties and attacks and under limited resources (such as the number of UAVs, communication data flow rates, and power consumptions).

A team of UAVs is subject constantly to uncertainties due to natural obstacles such as buildings, mountains, severe weather conditions that interrupt the network connections and observation capabilities, and to enemy attacks that may destroy some members and/or disrupt communications; see Figure 7.3. Consequently, the nodes and network topology switch during real-time operations.

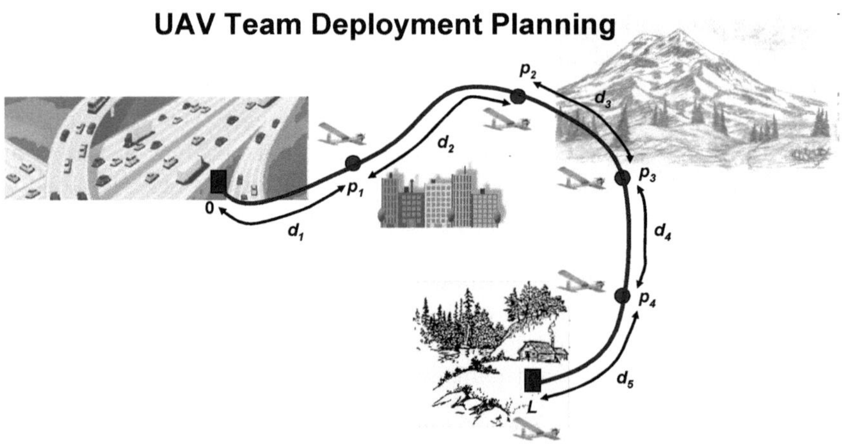

Fig. 7.1 UAVs on a deployment mission

This paper aims to introduce a new framework for UAV coordination and control, building on the emerging technology of network consensus control. The core target is to achieve suitable deployment of the team UAVs based on terrain conditions. In this paper, UAV deployment is formulated as a weighted and constrained consensus control problem that aims to coordinate all subsystems such that their formation converges to a desired distribution pattern. In UAV applications the desired pattern is that the weighted distances between consecutive UAVs are equal. Consensus control is an emerging

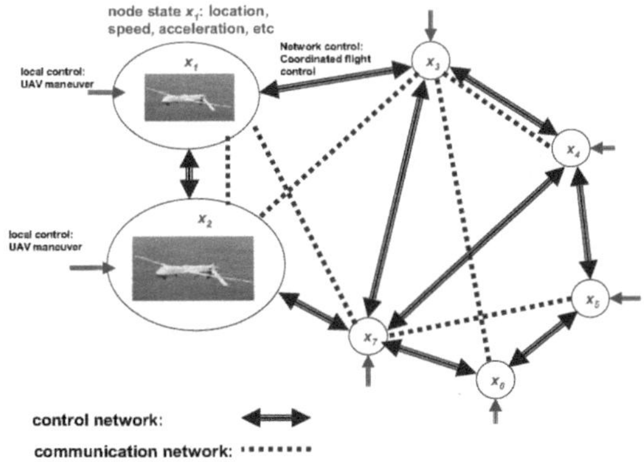

Fig. 7.2 Information and control network topology

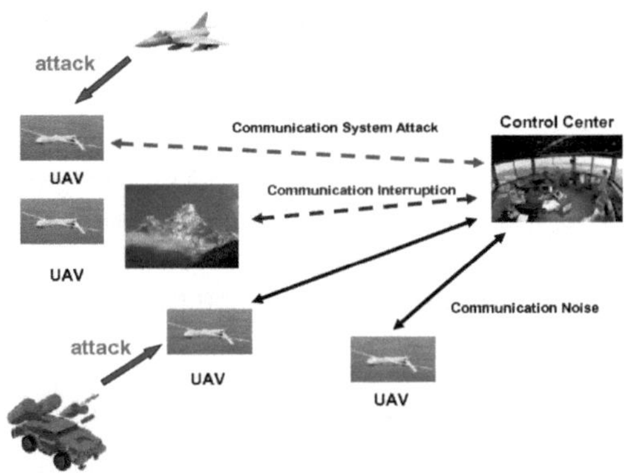

Fig. 7.3 Network Adversaries

field in networked control and remains an active research field. At present, most consensus controls are unconstrained and un-weighted. Nevertheless, UAV control is subject to terrain condition and deployment area constraints. As such, it commonly leads to a weighted and constrained consensus control problem.

Consensus control has drawn increased attention recently in a variety of application areas, including load balancing in parallel computing [21, 18], sensor networks [12], decentralized filtering, estimation, mobil agents [4], etc. The control methodologies developed up to date include deterministic control [4, 15], stochastic approximation algorithms [1, 2], switching network topologies [9, 11, 3, 5], etc.

Departing from the standard consensus control, the following four features are included in this paper.

1. In our recent work [23], a Markov model is used to treat a much larger class of systems, where the network graph is modulated by a discrete-time Markov chain. Our work in [23] also provides convergence and rates of convergence for the corresponding recursive algorithms.
2. Some of the useful features of [23] are extended to the weighted constrained consensus methodology in this paper.
3. In addition, the technique of post-iterate averaging is employed to enhance the power flow control algorithm. For detailed information on post-iterate averaging; see [13, 16] for its original introduction and [8, Chapter 11] for its extension to more general systems; see also [24]. With the iterate averaging, our algorithms provide the best convergence rate in terms of the best scaling factor and the smallest asymptotic covariance. Most significantly, they achieve asymptotically the well-known Cramér-Rao lower bounds [10], hence are best over all algorithms. This fast convergence feature is highly desirable for fast team formation.
4. Performance optimization is introduced to consensus control.

This framework offers several appealing features.

a. Local control to achieve a global deployment: Although a desired deployment is achieved for the entire team, each UAV only needs to communicate with its neighboring members as such communication costs and complexity remain minimal.
b. Scalability: Expanding and reduction of the team members does not complicate control strategies.
c. Robustness: Fluctuations in UAV positions, addition to or reduction of the UAVs can be readily accommodated. The weighted and constrained consensus control has applications in other areas such as platoon control of highway vehicles [20], power grid control [19], among others.

The rest of the paper is organized into the following sections. Section 7.2 describes how a typical UAV deployment problem can be formulated as a weighted and constrained consensus control problem. Algorithms for weighted and constrained consensus control are presented in Section 7.3. Their convergence properties and convergence rates are established. Section 7.4 further enhances the algorithms by post-iterate averaging. It is shown that consensus control for UAVs are subject to noises and their effect can be attenuated by the post-iterate averaging. Optimality of such modified algorithms in terms of convergence rates are established. Robustness and scalability of this framework are explained in Section 7.5 by showing its robustness against disturbances and adaptive capability when the network topology changes. Performance optimization is presented Section 7.6. It is shown that the control gain and step sizes can be selected separately. Consequently, the gain matrix can be selected to optimize a performance measure. The cases of optimal convergence rates and optimal robustness are presented. The feedback gains in consensus algorithms are selected to minimize a certain performance measure. Local implementation of such global optimization strategies is established. This leads a distributed optimization framework with the same network topology as the core consensus control. Finally, Section 7.7 points out future research directions.

7.2 UAV Coordinated Deployment Problems

7.2.1 Networked UAVs and Their Deployment

A team consists of r UAVs are to be deployed along a pathway of total length L. At time t, denote the total length of the surveillance range as $L(t)$. In the algorithm development, L is treated as a constant. Its changes will be viewed as a disturbance to the consensus control problem and d_i is the distance between UAV i and UAV $i-1$. We have the following constraint

$$\sum_{i=1}^{r} d_i(t) = L. \tag{7.1}$$

Due to terrain conditions, a desired coverage for an UAV differs at different locations. Each inter-vehicle distance has a terrain factor γ^i. The goal of the power flow control is to achieve consensus on weighted power d_i/γ^i, namely

$$\frac{d_i(t)}{\gamma^i} \to \beta, \quad i = 1, \dots, r$$

for some constant β. The convergence is either with probability one (w.p.1.) or in means squares (MS). For notational convenience in the algorithm development, we use $x^i(t) = P^i(t)$ and denote the state vector $x(t) = [P^1(t), \dots, P^r(t)]'$. The weighting coefficients are $\gamma = [\gamma^1, \dots, \gamma^r]'$, and the state scaling matrix $\Psi = \mathrm{diag}[1/\gamma^1, \dots, 1/\gamma^r]$, where v' is the transpose of a vector or a matrix v. Let $\mathbb{1}$ be the column vector of all 1s. Together with the constraint (7.1), the target of the constrained and weighted consensus control is

$$\Psi x(t) \to \beta \mathbb{1} \text{ subject to } \mathbb{1}' x(t) = L.$$

It follows from $\gamma' \Psi = \mathbb{1}'$ that

$$\beta = \frac{L}{\gamma' \mathbb{1}} = \frac{L}{\gamma^1 + \cdots + \gamma^r}.$$

The UAVs are linked by an information network, represented by a directed graph $\mathcal{G}$ whose element (i, j) (called a directed edge from node i to node j) indicates an observation between UAV i on the distance d_j. This network defines the information network: $(i, j) \in \mathcal{G}$ indicates estimation of the state d^j by UAV i via a communication link. Also, the factor γ^j is known. For node i, $(i, j) \in \mathcal{G}$ is a departing edge and $(l, i) \in \mathcal{G}$ is an entering edge. Due to the nature of power lines, we assume that if $(i, j) \in \mathcal{G}$ then $(j, i) \in \mathcal{G}$. The total number of communication links in $\mathcal{G}$ is l_s. From its physical meaning, node i can always observe its own state, which will not be considered as a link in $\mathcal{G}$.

For a selected time interval T, the consensus control is performed at the discrete-time steps $nT, n = 1, 2, \ldots$. At the control step n, the value of x will be denoted by $x_n = [x_n^1, \ldots, x_n^r]'$. Power flow control updates x_n to x_{n+1} by the amount u_n

$$x_{n+1} = x_n + u_n \tag{7.2}$$

with $u_n = [u_n^1, \ldots, u_n^r]'$. In UAV deployment, a distance change d_n^{ij} (called link control) from UAV i to UAV j at the nth step is the decision variable. The control u_n^i is determined by the link control d_n^{ij} as i is

$$u_n^i = - \sum_{(i,j)\in\mathcal{G}} d_n^{ij} + \sum_{(j,i)\in\mathcal{G}} d_n^{ji}. \tag{7.3}$$

The most relevant implication in this control scheme is that for all n,

$$\sum_{i=1}^{r} x_n^i = \sum_{i=1}^{r} x_0^i = L \tag{7.4}$$

that is, the constraint (7.1) is always satisfied. Consensus control seeks control algorithms such that $\Psi x_n \to \beta \mathbb{1}$ under the constraint (7.4).

A link $(i, j) \in \mathcal{G}$ entails an estimate $\widehat{x}_n^{ij}$ of x_n^j by node i with observation noise d_n^{ij}. That is,

$$\widehat{x}_n^{ij} = x_n^j + d_n^{ij}. \tag{7.5}$$

Let $\widetilde{x}_n$ and d_n be the l_s-dimensional vectors that contain all $\widehat{x}_n^{ij}$ and d_n^{ij} in a selected order, respectively. Then, (7.5) can be written as

$$\widetilde{x}_n = H_1 x_n + d_n \tag{7.6}$$

where H_1 is an $l_s \times r$ matrix whose rows are elementary vectors such that if the ℓth element of $\widetilde{x}_n$ is $\widehat{x}^{ij}$ then the ℓth row in H_1 is the row vector of all zeros except for a "1" at the jth position. Each link in $\mathcal{G}$ provides information $\delta_n^{ij} = x_n^i/\gamma^i - \widehat{x}_n^{ij}/\gamma^j$, an estimated difference between weighted x_n^i and x_n^j. This information may be represented by a vector δ_n of size l_s containing all δ_n^{ij} in the same order as $\widetilde{x}_n$. δ_n can be written as

$$\delta_n = H_2 \Psi x_n - \widetilde{\Psi} \widetilde{x}_n = H_2 \Psi x_n - \widetilde{\Psi} H_1 x_n - \widetilde{\Psi} d_n = H x_n - \widetilde{\Psi} d_n, \tag{7.7}$$

where the link scaling matrix $\widetilde{\Psi}$ is the $l_s \times l_s$ diagonal matrix whose k-th diagonal element is $1/\gamma^j$ if the k-th element of $\widetilde{x}_n$ is $\widehat{x}_n^{ij}$; H_2 is an $l_s \times r$ matrix whose rows are elementary vectors such that if the ℓth element of $\widetilde{x}(k)$ is $\widehat{x}^{ij}$ then the ℓth row in H_2 is the row vector of all zeros except for a "1" at the ith position, and $H = H_2\Psi - \widetilde{\Psi}H_1$.

Due to network constraints, the information δ_n^{ij} can only be used by nodes i and j. When the power control is linear, time invariant, and memoryless, we have $p_n^{ij} = \mu_n g_{ij}\delta_n^{ij}$ where g_{ij} is the link control gain and μ_n is a global time-varying scaling factor which will be used in state updating algorithms as the recursive step size. Let G be the $l_s \times l_s$ diagonal matrix that has g_{ij} as its diagonal element. In this case, the control becomes $u_n = -\mu_n J'G\delta_n$, where $J = H_2 - H_1$. For convergence analysis, we note that μ_n is a global control variable and we may represent u_n equivalently as

$$\begin{aligned} u_n &= -\mu_n J'G(Hx_n - \widetilde{\Psi}d_n) \\ &= -\mu_n(J'GHx_n - J'G\widetilde{\Psi}d_n) \\ &= \mu_n(Mx_n + Wd_n), \end{aligned} \tag{7.8}$$

with $M = -J'GH$ and $W = J'G\widetilde{\Psi}$. This, together with (7.2), leads to

$$x_{n+1} = x_n + \mu_n(Mx_n + Wd_n). \tag{7.9}$$

It can be directly verified that $\widetilde{\Psi}H_1\Psi^{-1} = H_1$, $H\Psi^{-1} = J$, $J\mathbb{1} = 0$, $\Psi^{-1}\mathbb{1} = \gamma$. These imply that $\mathbb{1}'M = 0$, $\mathbb{1}'W = 0$, $M\Psi^{-1}\mathbb{1} = M\gamma = 0$. The following assumption is imposed on the network.

(A0) The following conditions hold:

(1) All link gains are positive, $g_{ij} > 0$.
(2) $\mathcal{G}$ is strongly connected[1].

7.2.2 An Illustrative Example

We now use an example to illustrate the above concepts.

Example 1. A team of three UAVs must cover a surveillance line of length L, see Figure 7.4. UAV 1 controls the distance d_1, UAV 2 controls the distance d_2, and UAV 3 controls the distance d_3. Although d_4 is in fact a dependent variable since $d_1 + d_2 + d_3 + d_4 = L$, for systematic development, it is still formulated as a controlled variable. Then the condition $d_1 + d_2 + d_3 + d_4 = L$ is imposed as an additional constraint. The information topology is that in addition to observing their own controlled variables, UAV 1 observes also

[1] A directed graph is called strongly connected if there is a path from each node in the graph to every other node.

d_2, UAV 2 observes also d_1 and d_3, UAV 3 observes d_2 and d_4. the controller for d_4 obseerves d_3 also. The total length $L = 53.9$ km. Terrain factors $\gamma^1 = 12$, $\gamma^2 = 15$, $\gamma^3 = 20$, and $\gamma^4 = 28$. As a result,

$$\mathcal{G} = \{(1,2),(2,1),(2,3),(3,2),(3,4),(4,3)\}$$

$x = [d_1, d_2, d_3, d_4]'$, $\gamma = [12, 15, 20, 28]'$, $\Psi = \text{diag}[1/12, 1/15, 1/20, 1/28]$.

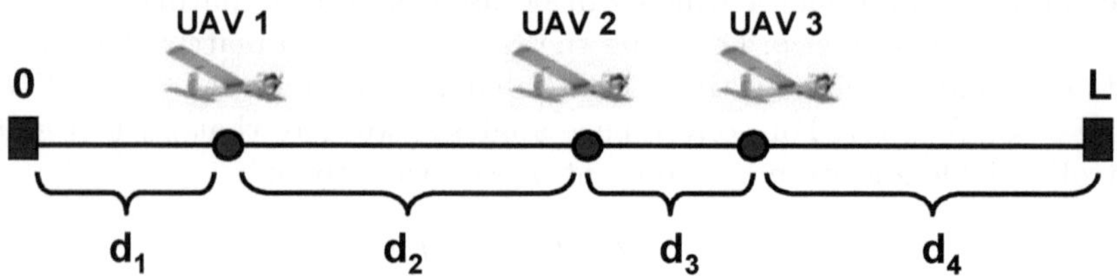

Fig. 7.4 A team of three UAVs

Since $L = 53.9$, we have

$$\beta = \frac{L}{\gamma^1 + \gamma^2 + \gamma^3 + \gamma^4} = 0.7187$$

and the weighted consensus is $\Psi x = 0.7187\mathbb{1}$ or

$$x = 0.7187\Psi^{-1}\mathbb{1} = [8.624, 10.781, 14.374, 20.124]'$$

By choosing the order for the links as $(1,2),(2,1),(2,3),(3,2),(3,4),(4,3)$, we have

$$\widetilde{x} = [\widehat{x}^{12}, \widehat{x}^{21}, \widehat{x}^{23}, \widehat{x}^{32}, \widehat{x}^{34}, \widehat{x}^{43}]'$$

and

$$H_1 = \begin{bmatrix} 0\,1\,0\,0 \\ 1\,0\,0\,0 \\ 0\,0\,1\,0 \\ 0\,1\,0\,0 \\ 0\,0\,0\,1 \\ 0\,0\,1\,0 \end{bmatrix}; H_2 = \begin{bmatrix} 1\,0\,0\,0 \\ 0\,1\,0\,0 \\ 0\,1\,0\,0 \\ 0\,0\,1\,0 \\ 0\,0\,1\,0 \\ 0\,0\,0\,1 \end{bmatrix}.$$

It follows that $\widetilde{\Psi} = \text{diag}[1/15, 1/12, 1/20, 1/15, 1/28, 1/20]$ and

$$\begin{aligned}
H &= H_2\Psi - \widetilde{\Psi}H_1 \\
&= \begin{bmatrix} 1/12 & 0 & 0 & 0 \\ 0 & 1/15 & 0 & 0 \\ 0 & 1/15 & 0 & 0 \\ 0 & 0 & 1/20 & 0 \\ 0 & 0 & 1/20 & 0 \\ 0 & 0 & 0 & 1/28 \end{bmatrix} - \begin{bmatrix} 0 & 1/15 & 0 & 0 \\ 1/12 & 0 & 0 & 0 \\ 0 & 0 & 1/20 & 0 \\ 0 & 1/15 & 0 & 0 \\ 0 & 0 & 0 & 1/28 \\ 0 & 0 & 1/20 & 0 \end{bmatrix} \\
&= \begin{bmatrix} 1/12 & -1/15 & 0 & 0 \\ -1/12 & 1/15 & 0 & 0 \\ 0 & 1/15 & -1/20 & 0 \\ 0 & -1/15 & 1/20 & 0 \\ 0 & 0 & 1/20 & -1/28 \\ 0 & 0 & -1/20 & 1/28 \end{bmatrix},
\end{aligned}$$

$$J = H_2 - H_1 = \begin{bmatrix} 1 & -1 & 0 & 0 \\ -1 & 1 & 0 & 0 \\ 0 & 1 & -1 & 0 \\ 0 & -1 & 1 & 0 \\ 0 & 0 & 1 & -1 \\ 0 & 0 & -1 & 1 \end{bmatrix}.$$

Suppose the control gains on the links are selected as $g_{12} = g_{21} = 3, g_{23} = g_{32} = 7$, $g_{34} = g_{43} = 9$. Then $G = \text{diag}[3, 3, 7, 7, 9, 9]$. It follows that

$$M = -J'GH = \begin{bmatrix} -1/2 & 1/2.5 & 0 & 0 \\ 1/2 & -2/1.5 & 7/10 & 0 \\ 0 & 7/7.5 & -4/2.5 & 9/14 \\ 0 & 0 & 9/10 & -9/14 \end{bmatrix},$$

$$W = J'G\widetilde{\Psi} = \begin{bmatrix} 1/5 & -1/4 & 0 & 0 & 0 & 0 \\ -1/5 & 1/4 & 7/20 & -7/15 & 0 & 0 \\ 0 & 0 & -7/20 & 7/15 & 9/28 & -9/20 \\ 0 & 0 & 0 & 0 & -9/28 & 9/20 \end{bmatrix}.$$

Since $\mathbb{1}'J' = \mathbb{1}'(H_2 - H_1)' = 0$, we have $\mathbb{1}'M = 0$ and $\mathbb{1}'W = 0$. We can show that under Assumption (A0), M has rank $r-1$ and is negative semi-definite. The proof uses similar ideas as in [23] and hence is omitted here. Recall that a square matrix $\widetilde{Q} = (\widetilde{q}_{ij})$ is a generator of a continuous-time Markov chain if $\widetilde{q}_{ij} \geq 0$ for all $i \neq j$ and $\sum_j \widetilde{q}_{ij} = 0$ for each i. Note that a generator of the associated continuous-time Markov chain is irreducible if the system of equations

$$\begin{cases} \nu\widetilde{Q} = 0, \\ \nu\mathbb{1} = C \end{cases} \tag{7.10}$$

for a given constant $C > 0$ has a unique solution, where $\nu = [\nu_1, \ldots, \nu_r] \in \mathbb{R}^{1\times r}$ with $\nu_i/C > 0$ for each $i = 1, \ldots, r$. When $C = 1$, ν is the associated stationary distribution. Consequently, under Assumption (A0), M is a generator of a continuous-time irreducible Markov chain.

7.3 Weighted Consensus Control with Linear Constraints

7.3.1 Algorithms

We begin by considering the state updating algorithm (7.9)

$$x_{n+1} = x_n + \mu_n M x_n + \mu_n W d_n, \tag{7.11}$$

together with the constraint

$$\mathbb{1}' x_n = L, \tag{7.12}$$

where $\{\mu_n\}$ is a sequence of stepsizes, M is a generator of a continuous-time Markov chain (hence $\mathbb{1}'M = 0$), $\{d_n\}$ is a noise sequence.

Since the algorithm (7.11) is a stochastic approximation procedure, we can use the general framework in Kushner and Yin [8] to analyze the asymptotic properties. Since $\mathbb{1}'M = 0$ and $\mathbb{1}'W = 0$, starting from the initial condition with $\mathbb{1}'x_0 = L$, the constraint $\mathbb{1}'x_n = L$ is always satisfied by the algorithm structure.

(A1)

1. The stepsize satisfies the following conditions: $\mu_n \ge 0$, $\mu_n \to 0$ as $n \to \infty$, and $\sum_n \mu_n = \infty$.
2. The noise $\{d_n\}$ is a stationary ϕ-mixing sequence such that $Ed_n = 0$, $E|d_n|^{2+\Delta} < \infty$ for some $\Delta > 0$, and that the mixing measure $\widetilde{\phi}_n$ satisfies

$$\sum_{k=0}^{\infty} \widetilde{\phi}_n^{\Delta/(1+\Delta)} < \infty, \tag{7.13}$$

where

$$\begin{aligned} \widetilde{\phi}_n &= \sup_{A\in\mathcal{F}^{n+m}} E^{(1+\Delta)/(2+\Delta)} |P(A|\mathcal{F}_m) - P(A)|^{(2+\Delta)/(1+\Delta)}, \\ \mathcal{F}_n &= \sigma\{d_k; k < n\}, \quad \mathcal{F}^n = \sigma\{d_k; k \ge n\}. \end{aligned}$$

Under Assumption (A0), M has an eigenvalue 0 of multiplicity 1 and all other eigenvalues are in the left complex plan (i.e., the real parts of the eigenvalues are negative). The null space of M is spanned by the vector $\gamma = [\gamma^1, \ldots, \gamma^r]'$.

Some commonly used stepsize sequences includes $\mu_n = a/n^\alpha$ for $1/2 < \alpha \le 1$. In such cases, $\sum_{n=1}^{\infty} \mu_n = \infty$ but $\sum_{n=1}^{\infty} \mu_n^2 < \infty$.

Note that ϕ-mixing sequences contain independent noises as a special case. However, they can represent a much larger class of noises to accommodate communication uncertainties such as signal interference, signal fading, latency, etc. As a consequence of (A1), the ϕ-mixing implies that the noise sequence $\{d_n\}$ is strongly ergodic [6, p. 488] in that for any m

$$\frac{1}{n} \sum_{j=m}^{m+n-1} d_j \to 0, \quad \text{w.p.1 as } n \to \infty. \tag{7.14}$$

7.3.2 Convergence of Algorithms

To study the convergence of the algorithm (7.11), we employ the stochastic approximation methods developed in [8]. Instead of working with the discrete-time iterations, we examine sequences defined in an appropriate function space. This will enable us to get a limit ordinary differential equation (ODE). The significance of the ODE is that the stationary point is exactly the true value of the desired weighted consensus. Then, convergence becomes a stability issue. We define

$$t_n = \sum_{j=0}^{n-1} \mu_j, \; m(t) = \max\{n : t_n \le t\}, \tag{7.15}$$

the piecewise constant interpolation $x^0(t) = x_n$ for $t \in [t_n, t_{n+1})$, and the shift sequence $x^n(t) = x^0(t + t_n)$. Due to the page limitation, we shall only outline the main steps involved in the proof. We can first derive a preliminary estimate on the second moments of x_n.

Lemma 1. *Under Assumption* (A1), *for any* $0 < T < \infty$,

$$\sup_{n \le m(T)} E|x_n|^2 \le K \text{ and } \sup_{0 \le t \le T} E|x^n(t)|^2 \le K, \tag{7.16}$$

for some $K > 0$, *where* $m(\cdot)$ *is defined in* (*7.15*).

Proof. We only indicate the main ideas and leave most of the details out. Concerning the first estimate, because of the boundedness of the second moment $E|d_n|^2$, the condition $\sum_{j=1}^{\infty} \mu_j^2 < \infty$, we can derive

$$E|x_n| \le K + K \sum_{j=1}^{n} \mu_j E|x_j|^2. \tag{7.17}$$

Here and henceforth, K is used as a generic positive constant, whose values may change for different usage. Applying the Grownwall's inequality to

(7.17), and then taking sup over $n \leq m(T)$, the first error bound is obtained. Likewise, we can obtain the second estimate. □

Theorem 1. *Under Assumption* (A1), *the iterates generated by the stochastic approximation algorithm (7.11) satisfies* $\Psi x_n \to \beta \mathbb{1}$ *w.p.1 as* $n \to \infty$.

Ideas of Proof. We only present the main ideas below. We show that $\{x^n(\cdot)\}$ is equicontinuous in the extended sense (see [8, p. 102] for a definition) w.p.1. To verify this, we note that by the argument in the first part of the proof in [22, Theorem 3.1],

$$\sum_{j=1}^{\infty} \mu_j d_j \text{ converges w.p.1.}$$

Define $\Phi^0(t) = \sum_{j=1}^{m(t)-1} \mu_j d_j$ and $\Phi^n(t) = \Phi^0(t_n + t)$, where $m(\cdot)$ is defined in (7.15). Then we can show that for each $T > 0$ and $\varepsilon > 0$, there is a $\delta > 0$ such that

$$\limsup_n \sup_{0 \leq |t-s| \leq \delta} |\Phi^n(t) - \Phi^n(s)| \leq \varepsilon \text{ w.p.1.}$$

The above estimates together with the form of the recursion imply that $x^n(\cdot)$ is equicontinuous in the extended sense. Next, we can extract a convergent subsequence, which will be denoted by $x^{n_\ell}(\cdot)$. Then the Arzela-Ascoli theorem concludes that $x^{n_\ell}(\cdot)$ converges to a function $x(\cdot)$ which is the unique solution (since the recursion is linear in x) of the ordinary differential equation (ODE)

$$\dot{x}(t) = Mx(t). \tag{7.18}$$

Moreover, from basic properties of Markov chains (see [25, Appendix A.1]), as $t \to \infty$, the solution $x(t)$ to (7.18) satisfies that $x(t)$ converges to the set Γ. That is, $\text{dist}(x(t), \Gamma) \to 0$ as $t \to \infty$, where $\text{dist}(\cdot, \cdot)$ is the usual distance function defined by $\text{dist}(x, \Gamma) = \inf_{y \in \Gamma} |x - y|$. Consequently, as $n \to \infty$ and $q(n_\ell) \to \infty$, $x^{n_\ell}(\cdot + q(n_\ell)) \to \Gamma$.

Furthermore, the algorithm (7.11) together with $x_n' \mathbb{1} = L$ leads to the desired weighted consensus. The equilibria of the limit ODE (7.18) and this constraint lead to the following system of equations

$$\begin{cases} Mx = 0 \\ \mathbb{1}' x = L. \end{cases} \tag{7.19}$$

The irreducibility of M then implies that (7.19) has a unique solution $x_* = \beta \Psi^{-1} \mathbb{1} = \beta \gamma$, which is precisely the weighted consensus.

Example 2. We now use the system in Example 1 to demonstrate the weighted consensus control. As in Example 1, the total distance is 53.9 km. Suppose that the initial distance distribution from the three UAVs are $d_0^1 = 12$ km; $d_0^2 = 14$ km; $d_0^3 = 10.9$ km; $d_0^4 = 17$ km. Weighted consensus for UAV control aims to distribute distances according to the terrain

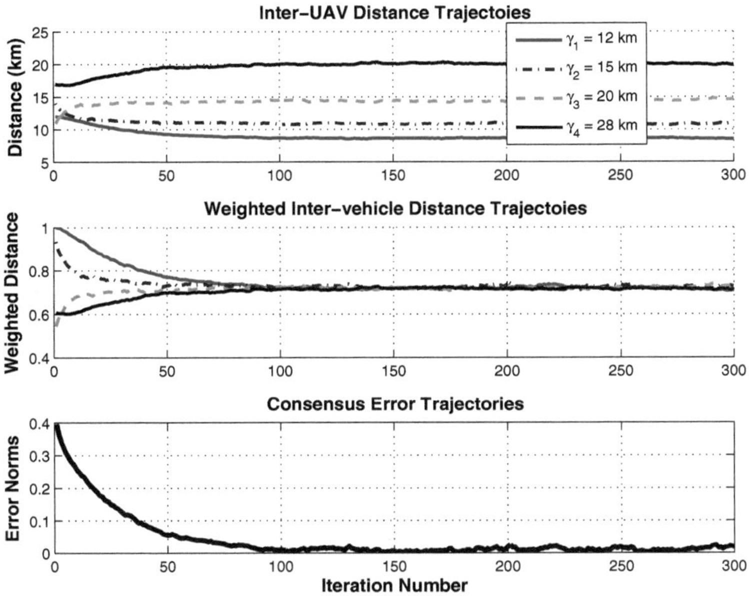

Fig. 7.5 UAV distance control with weighted consensus

conditions defined by $\gamma^1 = 12$, $\gamma^2 = 15$, $\gamma^3 = 20$, $\gamma^4 = 28$, with the total $\mathbb{1}'\gamma = 75$. The target percentage distance distribution over the whole length is $[12/75, 15/75, 20/75, 28/75] = [0.1600, 0.2000, 0.2667, 0.3733]$. From the total length of 53.9 km, the goal of weighted consensus is $d^1 = 8.624$ km; $d^2 = 10.780$ km; $d^3 = 14.373$ km; $d^4 = 20.123$ km.

Suppose that the link observation noises are i.i.d sequences of Gaussian noises with mean zero and variance 1. Figure 7.5 shows the inter-UAV distance trajectories. Staring from a large disparity in distance distribution, the top plot shows how distances are gradually distributed according to the terrain conditions. The middle plot illustrates that the weighted distances converge to a constant. The weighted consensus error trajectories are plotted in the bottom figure.

7.4 Post-Iterate Averaging for Improved Convergence under Large Observation Noise

The basic stochastic approximation algorithm (7.11) demonstrates desirable convergence properties under relatively small observation noises. However, its convergence rate is not optimal. Especially when noises are large, its convergence may not be sufficiently fast and its states show fluctuations. For example, for the same system as in Example 2, if the noise standard deviation is

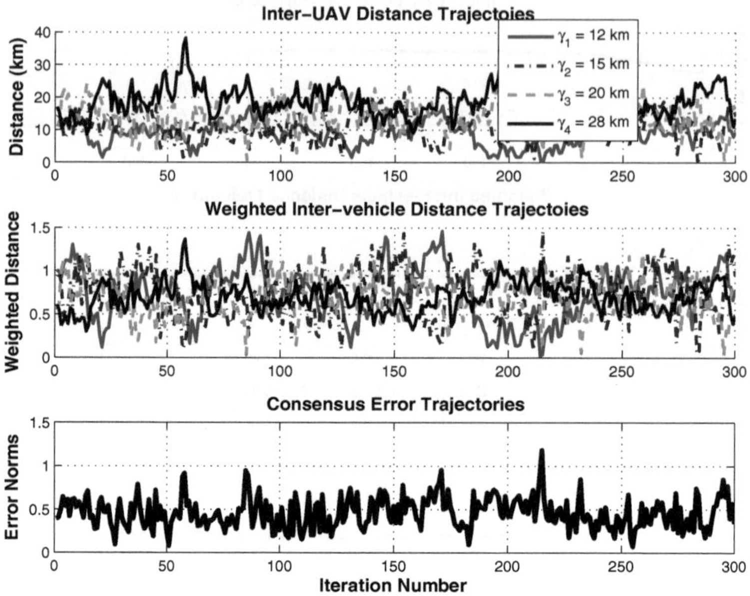

Fig. 7.6 UAV distance control with weighted consensus under large observation noise

increased from 1 to 20, its state trajectories demonstrate large variations, as shown in Figure 7.6.

To improve the efficiency, we take a post-iterate averaging, resulting in a two-stage stochastic approximation algorithm. For definiteness and simplicity, we take $\mu_n = c/n^\alpha$ for some $(1/2) < \alpha < 1$ and $c > 0$. The algorithm is modified to

$$\begin{aligned} x_{n+1} &= x_n + \frac{c}{n^\alpha} M x_n + \frac{c}{n^\alpha} W d_n \\ \overline{x}_{n+1} &= \overline{x}_n - \frac{1}{n+1}\overline{x}_n + \frac{1}{n+1} x_{n+1}. \end{aligned} \tag{7.20}$$

In what follows, for simplicity, we take $c = 1$ henceforth. Since $\mathbb{1}'M = 0$ and $\mathbb{1}'W = 0$, we have $\mathbb{1}'\overline{x}_n = L$. As a result, the constraint (7.1) remains satisfies after the post-iterate averaging. For some of the detailed analysis, we refer the reader to [24].

7.4.1 *Asymptotic Efficiency*

Strong convergence of the averaged $\overline{x}_n$ follows from that of x_n. This is stated in the following theorem with its proof omitted.

Theorem 2. *Suppose the conditions of Theorem 1 are satisfied. For iterates generated by algorithm (7.20) (together with the constraint* $\mathbb{1}' x_n = L$*),* $x_n \to \beta \Psi^{-1} \mathbb{1}$ *w.p.1 as* $n \to \infty$.

To proceed, we define

$$\overline{B}_n(t) = \frac{\lfloor nt \rfloor}{\sqrt{n}} (\Theta_{\lfloor nt \rfloor + 1} - \widetilde{x}_*). \tag{7.21}$$

We next show that asymptotically, the "effective" term of the normalized error above is given by $-\Gamma^{-1} B_n(t)$.

Lemma 2. *In addition to the assumptions* (A1)–(A3)*, assume* Γ *is a stable matrix (all of its eigenvalues have negative real parts). Then for* $t \in [0, 1]$,

$$\overline{B}_n(t) = -\Gamma^{-1} B_n(t) + o(1), \quad \textit{where} \quad o(1) \to 0$$

in probability uniformly in t *as* $n \to \infty$.

Remark 1. In the absence of the nonadditive noise, Γ becomes $\widetilde{M}$. The stability of $\widetilde{M}$ is verified by using the irreducibility of the generator M.

We are now ready to present the following theorem.

Theorem 3. *Under the conditions of Lemma 2, the following assertions hold:*

- $\overline{B}_n(\cdot)$ *converges weakly to* $\overline{B}(\cdot)$ *a Brownian motion whose covariance is given by* $\Gamma^{-1} \Sigma_0 (\Gamma^{-1})' t$;
- $\widetilde{x}_n - \widetilde{x}_*$ *is asymptotically normal with mean 0 and asymptotic covariance* $\Gamma^{-1} \Sigma_0 (\Gamma^{-1})'/n$.

Outline of Proof. To prove the first part of the theorem, we need only evaluate its covariance. This in turn follows from the well-known Slutsky theorem. To obtain the second part, set $t = 1$ in part one. Using Lemma 2 and part of the theorem, the desired result follows. □

We now establish the optimality of the algorithms. For clarity, we will include the dimension of the vector $\mathbb{1}$ in notation in the following derivations when needed. Also, the detailed proofs of the theorems are omitted. The reader is referred to our recent work [23] for details.

Partition the matrix M as

$$M = \begin{bmatrix} M_{11} & M_{12} \\ M_{21} & M_{22} \end{bmatrix}, \tag{7.22}$$

where $M_{11} \in \mathbb{R}^{(n-1)\times(n-1)}$, $M_{12} \in \mathbb{R}^{(n-1)\times 1}$, $M_{21} \in \mathbb{R}^{(n-1)\times 1}$, and $M_{22} \in \mathbb{R}^{1\times 1}$. Accordingly, we also partition $\bar{x}_n$, x_n, and W as

$$\overline{x}_n = \begin{bmatrix} \widetilde{\overline{x}}_n \\ \overline{x}^r_n \end{bmatrix}; \; x_n = \begin{bmatrix} \widetilde{x}_n \\ x^r_n \end{bmatrix}; \; W = \begin{bmatrix} \widetilde{W} \\ W_1 \end{bmatrix}, \tag{7.23}$$

respectively, with compatible dimensions with those of M.

Lemma 3. *Under Assumption A0, M_{11} is full rank.*

This result indicates that we can concentrate on $r-1$ components of $\overline{x}_n$. We can show that the asymptotic rate of convergence is independent of the choice of the $r-1$ state variables. To study the rates of convergence of $\overline{x}_n$, without loss of generality we need only examine that of $\widetilde{\overline{x}}_n$. It follows from that

$$\begin{cases} \widetilde{x}_{n+1} = \widetilde{x}_n + \mu_n(M_{11}\widetilde{x}_n + M_{12}x_n^r + \widetilde{W}d_n) \\ \qquad = \widetilde{x}_n + \mu_n(\widetilde{M}\widetilde{x}_n + \widetilde{W}d_n), \\ \widetilde{\overline{x}}_{n+1} = \widetilde{\overline{x}}_n - \dfrac{1}{n+1}\widetilde{\overline{x}}_n + \dfrac{1}{n+1}\widetilde{x}_{n+1}, \end{cases} \tag{7.24}$$

where

$$\widetilde{M} = M_{11} - M_{12}\mathbb{1}'_{r-1}.$$

Note that the noise is now reduced also to $\widetilde{W}d_n$, which is $r-1$ dimensional but is a function of l_s dimensional link noise d_n. Let $D = I_{r-1} + \mathbb{1}_{r-1}\mathbb{1}'_{r-1}$.

Lemma 4. *Assume* (A0). *Then $\widetilde{M} = M_{11}D$ and is full rank.*

For convergence speed analysis, let

$$e_n = \overline{x}_n - \beta\Psi^{-1}\mathbb{1}_n.$$

Decompose $e_n = [\widetilde{e}'_n, e^r_n]'$.

Theorem 4. *Suppose that $\{d_n\}$ is a sequence of i.i.d. random variables with mean zero and covariance $Ed_nd'_n = \Sigma$. Under Assumption* (A0), *the weighted consensus errors $\widetilde{e}_n$ satisfies that $\sqrt{n}\widetilde{e}_n$ converges in distribution to a normal random variable with mean* 0 *and covariance given by*

$$\widetilde{M}^{-1}\widetilde{W}\Sigma\widetilde{W}'(\widetilde{M}^{-1})'.$$

Note that the above result does not require any distributional information on the noise $\{\varepsilon(k)\}$ other than the zero mean and finite second moments. We now state the optimality of the algorithm when the density function is smooth.

Theorem 5. *Suppose that the noise $\{d_n\}$ is a sequence of i.i.d. noise with a density $f(\cdot)$ that is continuously differentiable. Then the recursive sequence $\widetilde{\overline{x}}_n$ is asymptotically efficient in the sense of the Cramér-Rao lower bound on $E\widetilde{e}'_n\widetilde{e}_n$ being asymptotically attained,*

$$nE\widetilde{e}'_n\widetilde{e}_n \to \mathrm{tr}(\widetilde{M}^{-1}\widetilde{W}\Sigma\widetilde{W}'(\widetilde{M}^{-1})'). \tag{7.25}$$

The convergence speed and optimality of e_n is directly related to these of $\widetilde{e}_n$.

Corollary 1. *Under the conditions of Theorem 5, the sequence $\{\overline{x}_n\}$ is asymptotically efficient in the sense of the Cramér-Rao lower bound on Ee'_ne_n*

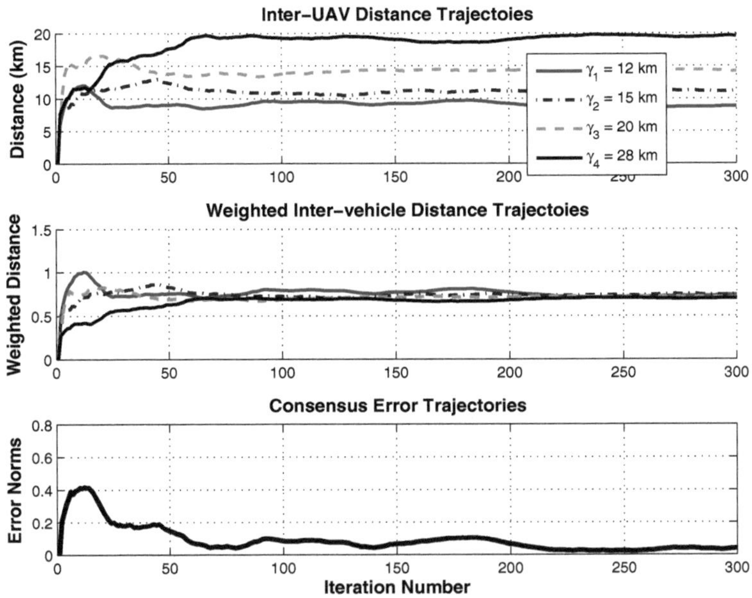

Fig. 7.7 UAV distance control with post-iterate averaging on weighted consensus algorithms

being asymptotically attained. The asymptotically optimal convergence speed is

$$nEe_n'e_n \to \mathrm{tr}(D\widetilde{M}^{-1}\widetilde{W}\Sigma\widetilde{W}'(\widetilde{M}^{-1})') \tag{7.26}$$

where $D = I_{r-1} + 1\!\!1_{r-1}1\!\!1_{r-1}'$.

Example 3. We now use the system in Example 2 to illustrate the effectiveness of post-iterate averaging. Suppose that the link observation noises are i.i.d sequences of Gaussian noises of mean zero and standard deviation 20. Now, the consensus control is expanded with post-iterate averaging. Figure 7.7 shows the distance trajectories. The distance distributions converge to the weighted consensus faster with much less fluctuations.

7.5 Robustness and Scalability

7.5.1 Robustness to Disruption of Terrain Conditions and Mission Goals

When a mission changes its tasks, say by modifying its length L of the surveillance range, they are represented by a sudden change in L. UAV coordination will re-distribute the inter-vehicle distances by the weighted consensus to reach a new equilibrium of consensus.

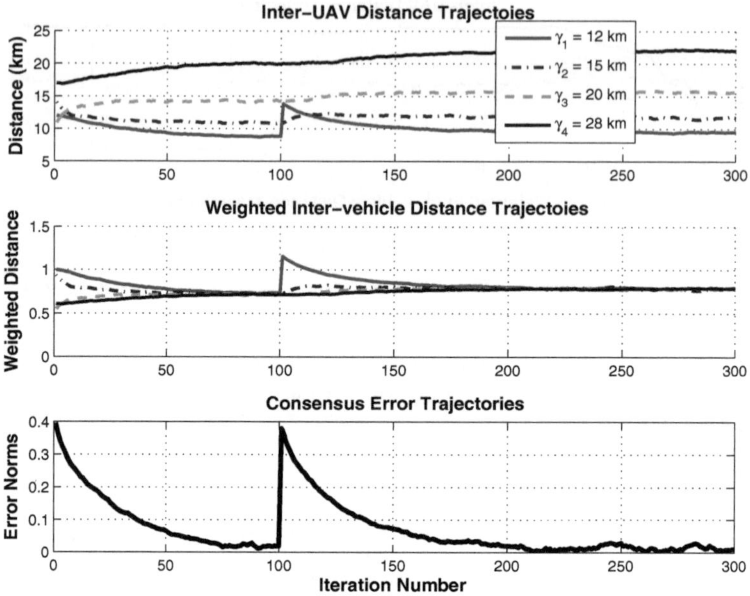

Fig. 7.8 Robustness of UAV distance control against terrain changes

Example 4. We now use the system in Example 2 to illustrate the effectiveness of robustness against load disruptions. Suppose that at the iteration step $n = 100$, a sudden increase of the total length by 5 km occurs to the surveillance mission. Consensus control then distributes it fairly according to the terrain conditions (represented by γ). Figure 7.8 shows the inter-UAV distance trajectories. The distance distributions converge to the weighted consensus.

7.5.2 Scalability

A team of UAVs often encounters dynamic changes in its team members or its information topology. When an UAV was damaged, it must retreat from the mission. Conversely, an enhancement of the team by additional UAVs changes the team composition and topologies. Both cases entail re-distribution of inter-UAV distances. In a centralized control scheme in which all information on UAVs is used by a central controller, the control strategy must be adapted for the entire system each time the UAV team topology changes. In our neighborhood-based network control method, an addition or deletion of an UAV will only affect its neighboring UAVs. In fact, all other UAVs will never be aware of changes in other parts of the team. However, by iterative control, an additional distance will be properly distributed throughout the entire team formation.

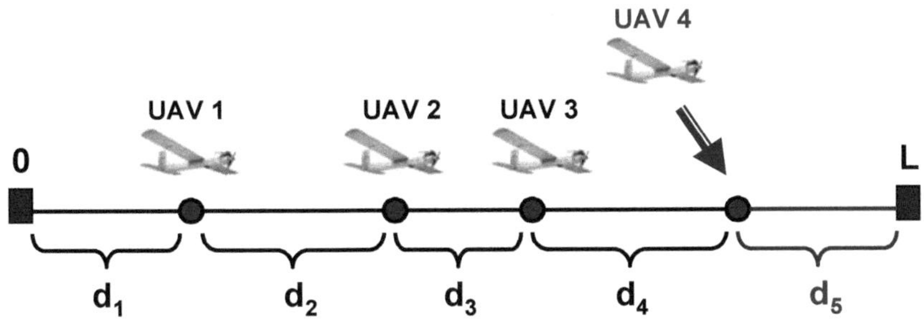

Fig. 7.9 An expanded UAV team from three members to four members

To be more concrete, suppose that a new UAV, labeled $r+1$, is added to the team with terrain factor γ^{r+1}. Assume that the new UAV is linked to UAV r only. Then, by (7.3), the control of UAV r will be modified from the original

$$x_{n+1}^r = x_n^r - \sum_{(r,j)\in\mathcal{G}} p_n^{rj} + \sum_{(j,r)\in\mathcal{G}} p_n^{jr}$$

to a slightly modified scheme

$$x_{n+1}^r = x_n^r - \sum_{(r,j)\in\mathcal{G}} p_n^{rj} + \sum_{(j,r)\in\mathcal{G}} p_n^{jr} - p_n^{r,r+1} + p_n^{r+1,r}$$

with all other bus control functions unchanged. The additional consumption of communication resources will be limited to the communication channel between UAV r and UAV $r+1$. This distributed and scalable control strategy is essential for UAV operations in reducing communication requirements and control complexity.

Example 5. Consider the same system as in Example 2. Suppose that at the iteration step $n = 100$, a new UAV becomes available. The new UAV 4 has weighting 10 and observes d_4 only. This addition results in a network topology change, leading to the new matrices derived below. The expanded network is shown in Figure 7.9. For systematic system analysis and presentation, we treat this at the system level with new M and W matrices. Note that for implementation of the consensus algorithms, only the control action of UAV 3 is affected with an additional term representing the link between D_4 and d_5.

Capacity factors are now expanded to $\gamma = [12, 15, 20, 28, 10]'$. The grid network set is expanded to

$$\mathcal{G} = \{(1,2), (2,1), (2,3), (3,2), (3,4), (4,3), (4,5), (5,4)\}.$$

The nework state vector is

$$x = [P^1, P^2, P^3, P^4, P^5]',$$
$$\Psi = \mathrm{diag}[1/120, 1/150, 1/200, 1/280, 1/100].$$

Since $L = 10 + 12 + 8.9 + 23 = 53.9$ is unchanged, we have the new weighted consensus

$$\beta = \frac{L}{\gamma^1 + \gamma^2 + \gamma^3 + \gamma^4 + \gamma^5} = 0.6341,$$

and the new weighted power distribution

$$x = 0.6341 \Psi^{-1} \mathbb{1} = [7.609, 9.512, 12.682, 17.755, 6.341]'.$$

By choosing the order for the links as $(1,2), (2,1), (2,3), (3,2), (3,4), (4,3), (4,5), (5,4)$, we have

$$\widetilde{x} = [\widehat{x}^{12}, \widehat{x}^{21}, \widehat{x}^{23}, \widehat{x}^{32}, \widehat{x}^{34}, \widehat{x}^{43}, \widehat{x}^{45}, \widehat{x}^{54}]'$$

and

$$H_1 = \begin{bmatrix} 0&1&0&0&0 \\ 1&0&0&0&0 \\ 0&0&1&0&0 \\ 0&1&0&0&0 \\ 0&0&0&1&0 \\ 0&0&1&0&0 \\ 0&0&0&0&1 \\ 0&0&0&1&0 \end{bmatrix}; H_2 = \begin{bmatrix} 1&0&0&0&0 \\ 0&1&0&0&0 \\ 0&1&0&0&0 \\ 0&0&1&0&0 \\ 0&0&1&0&0 \\ 0&0&0&1&0 \\ 0&0&0&1&0 \\ 0&0&0&0&1 \end{bmatrix}.$$

It follows that $\widetilde{\Psi} = \text{diag}[1/15, 1/12, 1/20, 1/15, 1/28, 1/20, 1/10, 1/28]$ and

$$\begin{aligned} H &= H_2 \Psi - \widetilde{\Psi} H_1 \\ &= \begin{bmatrix} 0.083 & -0.067 & 0 & 0 & 0 \\ -0.083 & 0.067 & 0 & 0 & 0 \\ 0 & 0.067 & -0.050 & 0 & 0 \\ 0 & -0.067 & 0.050 & 0 & 0 \\ 0 & 0 & 0.050 & -0.036 & 0 \\ 0 & 0 & -0.050 & 0.036 & 0 \\ 0 & 0 & 0 & 0.036 & -0.100 \\ 0 & 0 & 0 & -0.036 & 0.100 \end{bmatrix}, \end{aligned}$$

$$\begin{aligned} J &= H_2 - H_1 \\ &= \begin{bmatrix} 1 & -1 & 0 & 0 & 0 \\ -1 & 1 & 0 & 0 & 0 \\ 0 & 1 & -1 & 0 & 0 \\ 0 & -1 & 1 & 0 & 0 \\ 0 & 0 & 1 & -1 & 0 \\ 0 & 0 & -1 & 1 & 0 \\ 0 & 0 & 0 & 1 & -1 \\ 0 & 0 & 0 & -1 & 1 \end{bmatrix}. \end{aligned}$$

Suppose the control gains on the new link are $g_{45} = g_{54} = 10$. Then $G = \text{diag}[3, 3, 7, 7, 9, 9, 10, 10]$. It follows that

$$M = -J'GH = \begin{bmatrix} -0.5 & 0.4 & 0 & 0 & 0 \\ 0.5 & -1.333 & 0.7 & 0 & 0 \\ 0 & 0.933 & -1.6 & 0.643 & 0 \\ 0 & 0 & 0.9 & -1.357 & 2 \\ 0 & 0 & 0 & 0.714 & -2 \end{bmatrix}$$

$$W = J'G\widetilde{\Psi} = \begin{bmatrix} 0.2 & -0.25 & 0 & 0 & 0 & 0 & 0 & 0 \\ -0.2 & 0.25 & 0.35 & -0.467 & 0 & 0 & 0 & 0 \\ 0 & 0 & -0.35 & 0.467 & 0.321 & -0.45 & 0 & 0 \\ 0 & 0 & 0 & 0 & -0.321 & 0.45 & 1 & -0.357 \\ 0 & 0 & 0 & 0 & 0 & 0 & -1 & 0.357 \end{bmatrix}.$$

After the new UAV is added into the system, consensus control distributes inter-UAV distances according to the terrain factors of all distances. Figure 7.10 shows the distance trajectories. Initially, the new UAV reduces its neighbor's (UAV 4) distances d_4, due to its direct link to it. But, afterward the control adjusts distances among other UAVs throughout the entire team. The distance distributions converge to the new weighted consensus.

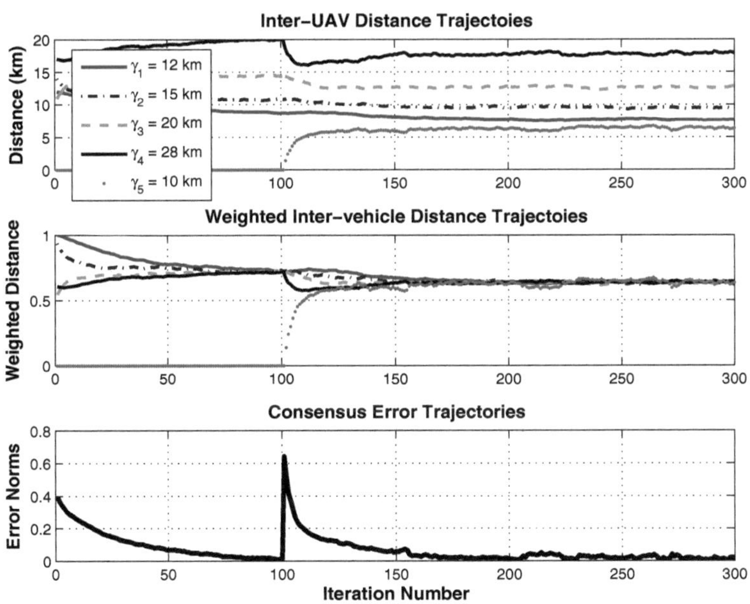

Fig. 7.10 UAV deployment network topology changes

7.6 Performance Optimization

7.6.1 Control Gains and Performance Optimization

Due to the C-R lower bound, by (7.26) the asymptotic estimation errors that can be achieved by any algorithm are explicitly given by

$$nEe_n'e_n \to \mathrm{tr}(D\widetilde{M}^{-1}\widetilde{W}\Sigma\widetilde{W}'(\widetilde{M}^{-1})').$$

In other words, the best convergence rates are in terms of mean-squares errors are $\mathrm{tr}(D\widetilde{M}^{-1}\widetilde{W}\Sigma\widetilde{W}'(\widetilde{M}^{-1})')/n$. Consequently, to improve convergence rates, it is desirable to reduce $\mathrm{tr}(D\widetilde{M}^{-1}\widetilde{W}\Sigma\widetilde{W}'(\widetilde{M}^{-1})')$.

Since $M = -J'GH$, $W = J'G\widetilde{\Psi}$, $D = I_{r-1} + \mathbb{1}_{r-1}\mathbb{1}_{r-1}'$,

$$M = \begin{bmatrix} M_{11} & M_{12} \\ M_{21} & M_{22} \end{bmatrix}, W = \begin{bmatrix} \widetilde{W} \\ W_{1}, \end{bmatrix}, \widetilde{M} = M_{11} - M_{12}\mathbb{1}_{r-1}',$$

it is clear that other than the structural variables that are determined by the network topology, the gain matrix G is the only design variable in this expression. By Assumption A0, G is a diagonal matrix with all diagonal elements positive. This section will explore optimization of G for different performance measures. In this section, we use the notation $G = \mathrm{diag}[g_i]$, $i = 1, \ldots, l_s$.

Denote $\eta(G) = \mathrm{tr}(D\widetilde{M}^{-1}\widetilde{W}\Sigma\widetilde{W}'(\widetilde{M}^{-1})')$. We note that η_{opt} is invariant under scaling of G. This is due to the fact that for any $c > 0$, $\eta(cG) = \eta(G)$. This observation point to a *design separation principle*: The gain matrix G and the step size μ_n in (7.9) can be designed separately. In other words, change in the step size μ_n will affect convergence but not the C-R lower bound which depends on G. As a result, in this section, we will concentrate on designing G to minimize certain performance measures.

7.6.2 Convergence Rate: Optimization

We start with design of G to minimize the MS estimation errors so that the convergence rate is optimized. The convergence rate optimization amounts to

$$\eta_{opt} = \min_{g_i > 0, i=1,\ldots,l_s} \eta(G). \tag{7.27}$$

Since for any $c > 0$, $\eta(cG) = \eta(G)$, in search of the optimal G, we may limit the search range to $1 \geq g_i > 0$,

$$\eta_{opt} = \min_{1 \geq g_i > 0, i=1,\ldots,l_s} \eta(G). \tag{7.28}$$

Example 6. Consider the same system as in Example 5. The selected gain matrix in Example 5 is $G = \mathrm{diag}[3, 3, 7, 7, 9, 9, 10, 10]$, which may be equivalent scaled to $0.05G = \text{diag}[0.15, 0.15, 0.35, 0.35, 0.45, 0.45, 0.5, 0.5]$ without affecting $\eta(G)$. This choice of G results in $\eta(G) = 1.5915$.

A Monte Carlo search is performed to find η_{opt} in (7.28). G is searched over 5000 randomly-selected sample points with

$$G = \text{diag}[\text{rand}, \text{rand}, \text{rand}, \text{rand}, \text{rand}, \text{rand}, \text{rand}, \text{rand}],$$

where rand is a random variable of uniform distribution in $(0, 1]$. The minimum $\eta(G)$ over this set is $\eta(G_0) = 1.3298$ which is achieved by

$$G_0 = \text{diag}[0.2078, 0.6030, 0.4909, 0.7039, 0.4047, 0.8076, 0.6913, 0.7920].$$

Due to performance optimization, this design exceeds the performance of the control in Example 5.

7.6.3 Min-Max Optimization

In UAV missions, communication systems are subject to uncertainties due to terrain conditions, inter-UAV distance changes, and adversary signal jammer. Consequently, accurate channel noise characterizations are often difficult. This implies that the matrix Σ in (7.26) may not be known. Here we assume that Σ belongs to a set Ω of potential noise characterizations. In this case, we should use $\eta(G, \Sigma)$ to indicate its dependence on Σ. To ensure reliable control performance, we seek the optimal gain matrix design under a worst-case scenario. Mathematically, this means the following min-max design problem

$$\eta_{opt} = \min_{1 \geq g_i > 0, i=1,\ldots,l_s} \max_{\Sigma \in \Omega} \eta(G, \Sigma). \tag{7.29}$$

Example 7. Consider the system in Example 6. In addition, we assume that one adversary signal jammer with unknown location can corrupt one communication, resulting in a substantially increased noise level from variance 1 to variance 100. In this case, Ω contains 8 possible values of Σ_i, $i = 1, \ldots, 8$ such that Σ_i is the diagonal matrix will all diagonal elements equal to 1,, except a 100 at the ith position.

A Monte Carlo search is performed to find η_{opt} in (7.29). G is searched over 5000 sample points with

$$G = \text{diag}[\text{rand}, \text{rand}, \text{rand}, \text{rand}, \text{rand}, \text{rand}, \text{rand}, \text{rand}],$$

where *rand* is a random variable of uniform distribution in $(0, 1]$. The worst-case

$$\max_{\Sigma_i, i=1,\ldots,8} \eta(G, \Sigma)$$

is calculated for each G. The minimum $\eta(G)$ over this set is $\eta(G_0) = 25.9553$ which is achieved by

$$G_0 = \text{diag}[0.5418, 0.0213, 0.7086, 0.4075, 0.8299, 0.7014, 0.8112, 0.0561].$$

It is observed that due to the worst-case situation, the control design is conservative and the achievable performance is significantly affected. This is to be expected in such a robust control design.

7.7 Concluding Remarks

This paper introduces a new control methodology for UAV coordination. The methodology is based on weighted and constrained consensus control that can take into considerations of terrain conditions. The scalability of the framework permits dynamic expansion of team UAVs without increasing communication, control, and computation complexities of network control functions.

This paper is a first attempt in this new direction. We have left many open issues. For instance, this paper considers communication uncertainty in terms of additive noises. Other types of communication uncertainties such as latency, gains, quantization errors, data compression, and packet losses, will be of interests in development of UAV technology. Integration of consensus control with dynamic control of subsystems, fault detection, dynamic stability of UAV flight control will be of values in gaining better control strategies for UAV coordination.

References

1. Cortes, J., Bullo, F.: Coordination and geometric optimization via distributed dynamical systems. SIAM J. Control Optim. (5), 1543–1574 (May 2005)
2. Huang, M., Manton, J.H.: Coordination and consensus of networked agents with noisy measurements: stochastic algorithms and asymptotic behavior. SIAM J. Control Optim. 48(1), 134–161 (2009)
3. Huang, M., Dey, S., Nair, G.N., Manton, J.H.: Stochastic consensus over noisy networks with Markovian and arbitrary switches. Automatica 46, 1571–1583 (2010)
4. Jadbabaie, A., Lin, J., Morse, A.S.: Coordination of groups of mobile autonomous agents using nearest neighbor rules. IEEE Trans. Automat. Contr. 48, 988–1000 (2003)
5. Kar, S., Moura, J.M.F.: Distributed consensus algorithms in sensor networks with imperfect communication: Link failures and channel noise. IEEE Trans. Signal Processing 57(1), 355–369 (2009)
6. Karlin, S., Taylor, H.M.: A First Course in Stochastic Processes, 2nd edn. Academic Press, New York (1975)
7. Kothari, D.P., Nagrath, I.J.: Modern Power System Analysis. McGraw Hill Higher Education (2008)
8. Kushner, H.J., Yin, G.: Stochastic Approximation and Recursive Algorithms and Applications, 2nd edn. Springer, New York (2003)

9. Moreau, L.: Stability of multiagent systems with time- dependent communication links. IEEE Trans. Autom. Control 50(2), 169–182 (2005)
10. Nelson, M.B., Khasminskii, R.Z.: Stochastic Approximation and Recursive Estimation. Amer. Math. Soc., Providence (1976)
11. Olfati-Saber, R., Fax, J.A., Murray, R.M.: Consensus and cooperation in networked multi-agent systems. IEEE Proc. 95(1), 215–233 (2007)
12. Ogren, P., Fiorelli, E., Leonard, N.E.: Cooperative control of mobile sensor networks: Adaptive gradient climbing in a distributed environment. IEEE Trans. Autom. Control 49(8), 1292–1302 (2005)
13. Polyak, B.T.: New method of stochastic approximation type. Automation Remote Control 7, 937–946 (1991)
14. Rau, N.S., Wan, Y.-H.: Optimum location of resources in distributed planning. IEEE Trans. Power Syst. 9, 2014–2020 (1994)
15. Ren, W., Beard, R.W.: Consensus seeking in multiagent systems under dynamically changing interaction topologies. IEEE Trans. Automat. Control 50(5), 655–661 (2005)
16. Ruppert, D.: Stochastic approximation. In: Ghosh, B.K., Sen, P.K. (eds.) Handbook in Sequential Analysis, pp. 503–529. Marcel Dekker, New York (1991)
17. Tenti, P., Paredes, H.K.M., Mattavelli, P.: Conservative Power Theory, a Framework to Approach Control and Accountability Issues in Smart Microgrids. IEEE Transactions on Power Electronics 26(3), 664–673 (2011)
18. Tsitsiklis, J.N., Bertsekas, D.P., Athans, M.: Distributed asynchronous deterministic and stochastic gradient optimization algorithms. IEEE Trans. Automat. Control 31(9), 803–812 (1986)
19. Wang, L.Y., Wang, C., Yin, G.: Weighted and Constrained Consensus for Distributed Power Flow Control. In: PMAPS 2012, Istanbul, Turkey, June 10-14 (2012)
20. Wang, L.Y., Syed, A., Yin, G., Pandya, A., Zhang, H.: Control of Vehicle Platoons for Highway Safety and Efficient Utility: Consensus with Switching Communication Network Topologies and Vehicle Dynamics (submitted)
21. Xiao, L., Boyd, S., Kim, S.J.: Distributed average consensus with least-mean-square deviation. Journal of Parallel and Distributed Computing 67, 33–46 (2007)
22. Yin, G.: On extensions of Polyak's averaging approach to stochastic approximation. Stochastics Stochastics Rep. 36, 245–264 (1991)
23. Yin, G., Sun, Y., Wang, L.Y.: Asymptotic properties of consensus-type algorithms for networked systems with regime-switching topologies. Automatica 47, 1366–1378 (2011)
24. Yin, G., Wang, L.Y., Sun, Y., Casbeer, D., Holsapple, R., Kingston, D.: Iterate averaging for asymptotic optimality in stochastic approximation algorithms of consensus type. To Appear in J. Control Theory Appl.
25. Yin, G., Zhang, Q.: Continuous-Time Markov Chains and Applications: A Singular Perturbation Approach. Springer, New York (1998)
26. Zhang, P., Guo, Y.: A novel spectral partitioning method for large-scale distribution system feeder reconfiguration. Automation of Electric Power Systems 26(18), 25–29 (2002)

9. Moreau, L.: Stability of multiagent systems with time-dependent communication links. IEEE Trans. Autom. Control 50(2), 169–182 (2005)
10. [illegible]: Stochastic Approximation and Recursive Estimation. Amer. Math. Soc., Providence (1976)
11. Olfati-Saber, R., Fax, J.A., Murray, R.M.: Consensus and cooperation in networked multi-agent systems. [illegible] (2007)
12. Ögren, P., Fiorelli, E., Leonard, N.E.: Cooperative control of mobile sensor networks: [illegible] distributed environment. IEEE Trans. Autom. Control 49(8), 1292–1302 (2004)
13. [illegible]
14. [illegible]
15. [illegible]

Author Index

MIX
Papier aus verantwortungsvollen Quellen
Paper from responsible sources
FSC® C105338

If you have any concerns about our products,
you can contact us on
ProductSafety@springernature.com

In case Publisher is established outside the EU,
the EU authorized representative is:
Springer Nature Customer Service Center GmbH
Europaplatz 3, 69115 Heidelberg, Germany

Printed by Libri Plureos GmbH
in Hamburg, Germany